MANAGING THE RISKS OF ORGANIZATIONAL ACCIDENTS

This book is dedicated to two pilots and two surgeons who have greatly enhanced the safety of their respective domains:

Captain Gordon Vette

Captain Daniel Maurino

Dr Lucian Leape

Mr Marc de Leval

MANAGING THE RISKS OF ORGANIZATIONAL ACCIDENTS

JAMES REASON

ASHGATE

Published by
Ashgate Publishing Limited
Gower House
Croft Road
Aldershot
Hants GU11 3HR
England

Ashgate Publishing Company
Suite 420, 101 Cherry Street,
Burlington, VT 05401-4405 USA

Ashgate website:http://www.ashgate.com

Reprinted 1998, 1999, 2000 (twice), 2001, 2002, 2003, 2004, 2005, 2006, 2008

British Library Cataloguing in Publication Data
Reason, James
  Managing the risks of organizational accidents
  1.Industrial accidents 2. Hazardous substances – Safety
  measures 3. Industrial safety – Management
  I. Title
  363.1´1´06

Library of Congress Cataloging-in-Publication Data
Reason, J. T.
    Managing the risks of organizational accidents / James Reason.
        p.   cm.
    ISBN 1 84014 105 0 (pbk)   1 84014 104 2 (hbk)
    1. Industrial accidents.   2. Risk assessment. I. Title.
  T54.R4   1997
  658.3´82—dc21   97–24648
CIP

ISBN 13: 978 1 84014 104 7 (Hbk)
ISBN 13: 978 1 84014 105 4 (Pbk)

Typeset by Manton Typesetters, 5–7 Eastfield Road, Louth, Lincolnshire, UK.
Printed in Great Britain by MPG Books Ltd, Bodmin, Cornwall

Contents

List of Figures

List of Tables

List of Abbreviations

ACAA   Australian Civil Aviation Authority
ALARP   as low as reasonably practicable
AMMS   Aurora Mishap Management System
AOC   Air Operators Certificate
ASRS   Aviation Safety Report System (NASA)
BASI   Bureau of Air Safety Investigation (Australia)
BASIS   British Airways Safety Information System
BB & Co.   Barings Brothers & Co.
BFS   Barings Futures (Singapore) Pte Limited
BSL   Barings Securities Limited
CEO   chief executive officer
CIMAH   Control of Industrial Major Hazards
COSHH   Control of Substances Hazardous to Health
CRIEPI   Central Research Institute for the Electrical Power Industry
CRM   crew (cockpit) resource management
EC   European Commission
EM   error management
EPC   error-producing condition
FAA   Federal Aviation Administration (US)
FDR   flight data recorder
FEA   failure mode and effects analysis
FMS   flight management system
FSA   formal safety assessment
GFT   general failure type
HAZAN   hazards operability study
HAZOP   hazard and operability study
HEA   human error analysis
HEART   Human Error Assessment and Reduction Technique
HEMP   Hazardous Effects Management Process
HRA   human reliability analysis
HRO   high-reliability organization
HSC   Health and Safety Commission
HSE   Health and Safety Executive
IAEA   International Safety Advisory Group

IDA	Influence Diagram Approach
IFSD	inflight engine shutdown
INPO	Institute of Nuclear Power Operations (US)
JAL	Japan Airlines
KB	knowledge-based
LII	lost-time injury
MEDA	Maintenance Error Decision Aid
MESH	Managing Engineering Safety Health
MSA	Marine Safety Agency
NASA	National Aeronautics and Space Administration
NCO	non-commissioned officer
NRC	Nuclear Regulatory Commission
NTSB	National Transport Safety Board
NUREG	Report series issued by Nuclear Regulatory Commission
NWA	Northwest Airlines
O & M	organizational and managerial
PIF	performance-influencing factor
PRA	probabilistic risk assessment
PSA	probabilistic safety assessment
PWR	pressurised water reactor
RAMS	reliability and maintainability study
RB	rule-based
RPF	railway problem factor
RBMK	A Soviet-built nuclear power plant
SB	skill-based
SESMA	Special Event Search and Master Analysis
SIMEX	Singapore Monetary Exchange
SOP	standard operating procedure
SPC	Statistical Process Control
SR & QA	Safety Reliability and Quality Assurance Program
TBR	to-be-remembered
TMI	Three Mile Island
TQM	Total Quality Management
VPC	violation-producing factor

Preface

This book is not meant for an academic readership, although I hope that academics and students might read it. It is aimed at 'real people' and especially those whose daily business is to think about, and manage or regulate, the risks of hazardous technologies. My imagined reader is someone with a technical background rather than one in human factors. To this end, I have tried—not always successfully—to keep the writing as jargon-free as possible.

The book is not targeted at any one domain. Rather, it tries to identify general principles and tools that are applicable to all organizations facing dangers of one sort or another. This includes banks and insurance companies just as much as nuclear power plants, oil exploration and production, chemical process plants and air, sea and rail transport. The more one moves towards the upper reaches of such systems, the more similar their organizational processes—and weaknesses—become.

In a book of this type the 'big bang' examples inevitably tend to predominate, but, although I have used case study examples to illustrate points, this is not intended to be yet another catalogue of accident case studies. My emphasis is upon principles and practicalities—the two must work hand-in-hand. But the real test is whether or not these ideas can eventually be translated into some improvement in the resistance of complex, well defended systems to rare, but usually catastrophic, 'organizational accidents'.

James Reason

1 Hazards, Defences and Losses

Individual and Organizational Accidents

There are two kinds of accidents: those that happen to individuals and those that happen to organizations. Individual accidents are by far the larger in number, but they are not the main concern of this book. Our focus will be upon *organizational accidents*. These are the comparatively rare, but often catastrophic, events that occur within complex modern technologies such as nuclear power plants, commercial aviation, the petrochemical industry, chemical process plants, marine and rail transport, banks and stadiums.

Organizational accidents have multiple causes involving many people operating at different levels of their respective companies. By contrast, individual accidents are ones in which a specific person or group is often both the agent and the victim of the accident.[1] The consequences to the people concerned may be great, but their spread is limited. Organizational accidents, on the other hand, can have devastating effects on uninvolved populations, assets and the environment. Whereas the nature (though not necessarily the frequency) of individual accidents has remained relatively unchanged over the years, organizational accidents are a product of recent times or, more specifically, a product of technological innovations which have radically altered the relationship between systems and their human elements.

Finding the Right Level of Explanation

Organizational accidents are difficult events to understand and control. They occur very rarely and are hard to predict or foresee. To the people on the spot, they happen 'out of the blue'. Difficult though they may be to model, we have to struggle to find some way of understanding the development of organizational accidents if we are

1

to achieve any further gains in limiting their occurrence. Quite apart from the human costs in deaths and injuries, there are very few commercial organizations that can survive the fallout from a major accident of this kind.

It has been said that nothing in logic is accidental. But does the reverse hold true? Is there nothing logical about accidents? Are there no underlying principles of accident causation? This book is written in the belief that such principles do exist. Organizational accidents may be truly accidental in the way in which the various contributing factors combine to cause the bad outcome, but there is nothing accidental about the existence of these precursors, nor in the conditions that created them. The difficulty, however, lies in finding the appropriate level of description.

If we consider only their surface details—the kind of information that is reported in press accounts—organizational accidents are dauntingly diverse. They involve a variety of systems in widely differing locations. Each accident has its own very individual pattern of cause and effect. Apart from the fact that they are all bad news, this level of description seems to defy generalization and implies that we clearly need to investigate more deeply into some common underlying structure and process to find the right level of explanation.

At the other extreme, it can be claimed that all organizational accidents involve the unplanned release of destructive agencies such as mass, energy, chemicals and the like. This is indeed a generalization, but it does not take us very far. However, like gunners, we have bracketed the target. The appropriate level of understanding has to lie somewhere between the highly idiosyncratic superficial details and the vagueness of this overly broad definition.

The aim is to find ideas that could be applied equally well to a wide range of low-risk, high-hazard domains. The basic thesis of this book is that the framework illustrated in Figure 1.1 will serve this purpose well. Figure 1.1 shows the relationship between the three elements that make up the title of this chapter: hazards, defences and losses. All organizational accidents entail the breaching of the barriers and safeguards that separate damaging and injurious hazards from vulnerable people or assets—collectively termed 'losses'. This is in sharp contrast to individual accidents where such defences are often either inadequate or lacking.

Figure 1.1 directs our attention to the central question in all accident investigation: By what means are the defences breached? Three sets of factors are likely to be implicated—human, technical and organizational—and all three will be governed by two processes common to all technological organizations: production and protection.

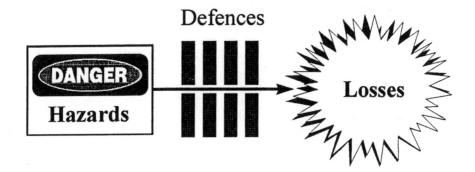

Figure 1.1   The relationship between hazards, defences and losses

Production and Protection: Two Universals

All technological organizations produce something—manufactured goods, the transportation of people, financial or other services, the extraction of raw materials and the like. But, to the extent that productive operations expose people and assets to danger, all organizations (and the larger systems within which they are embedded) require various forms of protection to intervene between the local hazards and their possible victims and lost assets.

While the productive aspects of an organization are fairly well understood and their associated processes relatively transparent, the protective functions are both more varied and more subtle. Figure 1.2 introduces some of the issues involved in the complex relationship between production and protection. In an ideal world, the level of protection should match the hazards of the productive operations—the parity zone.[2] The more extensive the productive operations, the greater is the hazard exposure and so also is the need for corresponding protection. But different types of production—and hence different organizations—vary in the severity of their operational hazards. Thus, low-hazard ventures will require less protection per productive unit than will high-hazard ventures. In other words, the former can operate in the region below the parity zone, whereas the latter must operate above it.

This broad operating zone (the lightly shaded area in Figure 1.2) is bounded by two dangerous extremes. In the top left-hand corner lies the region in which the protection far exceeds the dangers posed by the productive hazards. Since protection consumes productive resources—such as people, money and materials—such grossly overprotected organizations would probably soon go out of business.

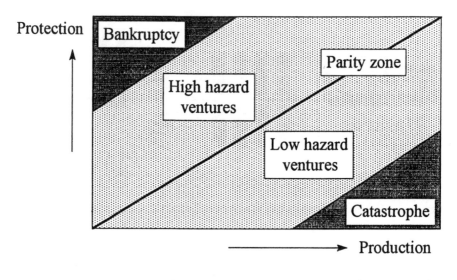

Figure 1.2   Outline of the relationship between production and protection

At the other extreme, in the bottom right-hand corner, the available protection falls far short of that needed for productive safety, and organizations operating in this zone face a very high risk of suffering a catastrophic accident (which probably also means going out of business). These obviously dangerous zones are generally avoided, if only because they are unacceptable to both the regulators and the shareholders. Our main concern is with how organizations navigate the space bounded by these two extremes.

Despite frequent protestations to the contrary, the partnership between production and protection is rarely equal, and one of these processes will predominate, depending on the local circumstances. Since production creates the resources that make protection possible, its needs will generally have priority throughout most of an organization's lifetime. This is partly because those who manage the organization possess productive rather than protective skills, and partly because the information relating to production is direct, continuous and readily understood. By contrast, successful protection is indicated by the absence of negative outcomes. The associated information is indirect and discontinuous. The measures involved are hard to interpret and often misleading. It is only after a bad accident or a frightening near-miss that protection comes—for a short period—uppermost in the minds of those who manage an organization.

All rational managers accept the need for some degree of protection. Many are committed to the view that production and protection necessarily go hand-in-hand in the long term. It is in the short term that conflicts occur. Almost every day, line managers and supervisors have to choose whether or not to cut safety corners in order to meet deadlines or other operational demands. For the most part, such short-cuts bring no bad effects and so can become an habitual part of routine work practices. Unfortunately, this gradual reduction in the system's safety margins renders it increasingly vulnerable to particular combinations of accident-causing factors.

Figure 1.3—the main purpose of which is to introduce the two important features of organizational life described below—plots the unhappy progress of one hypothetical organization through the production–protection space. The history starts towards the bottom left-hand corner of the space where the organization begins production with a reasonable safety margin. (The organization's progress between events is indicated by the black dots.) As time passes, the safety margin is gradually diminished until a low-cost accident occurs. The event leads to an improvement in protection, but this is then traded off for productive advantage until another, more serious, accident occurs. Again, the level of protection is increased, but this is gradually eroded by an event-free period. The life history ends with a catastrophe.

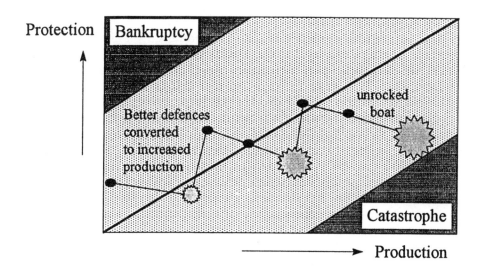

Figure 1.3 **The lifespan of a hypothetical organization through the production–protection space**

Trading off Added Protection for Improved Production

Improvements in protection are often put in place during the period immediately following a bad event. Although the aim is to avoid a repetition of an accident, it is soon appreciated that the improved defences confer productive advantages. Mine owners in the early nineteenth century, for example, quickly realized that the invention of the Davy lamp permitted coal to be extracted from areas previously considered too dangerous because of the presence of combustible gases. Ship owners soon discovered that marine radar allowed their merchant vessels to travel at greater speed through crowded or confined seaways. In short, protective gains are frequently converted into productive advantages, leaving the organization with the same inadequate protection that prevailed before the event or with something even worse. The incidence of mine explosions increased dramatically in the years following the introduction of the Davy lamp, and the history of marine accidents is littered with radar-assisted collisions—to name but two of the many examples of accidents brought about by sacrificing protective benefits for productive gains. This process has been termed 'risk compensation' or 'risk homeostasis'.[3]

The Dangers of the 'Unrocked Boat'[4]

There is plentiful evidence to show that a lengthy period without a serious accident can lead to the steady erosion of protection as productive demands gain the upper hand in this already unequal relationship. It is easy to forget to fear things that rarely happen, particularly in the face of productive imperatives such as growth, profit and market share. As a result, investment in more effective protection falls off and the care and maintenance necessary to preserve the integrity of existing defences declines. Furthermore, productive growth is regarded as commercially essential in most organizations. Simply increasing production without the corresponding provision of new or extended defences will also erode the available safety margins. The consequence of both processes—neglecting existing defences and failing to provide new ones—is a much increased risk of a catastrophic, and sometimes terminal, accident.

We will return to the interplay between production and protection later, but for now we need to focus on protection—the layers of defences, barriers and safeguards that are erected to withstand both natural and manmade hazards. The one sure fact about an accident is that the defences must have been breached or bypassed. Identifying how these breakdowns can occur is the first step in understanding the processes common to all organizational accidents.

Just as production can involve many different activities, so protection can be achieved in a variety of ways. In the remainder of this book, we will reserve the term 'protection' for the general goal of ensuring the safety of people and assets, and we will use the term 'defences' to refer to the various means by which this goal can be achieved. At this point it would be convenient to focus on the various ways by which defences may be described or classified.

The Nature and Variety of Defences

Defences can be categorized both according to the various functions they serve and by the ways in which these functions are achieved. Although defensive functions are universals, their modes of application will vary between organizations, depending on their operating hazards.

All defences are designed to serve one or more of the following functions:

- to create *understanding* and *awareness* of the local hazards
- to give clear *guidance* on how to operate safely
- to provide *alarms and warnings* when danger is imminent
- to *restore* the system to a safe state in an off-normal situation
- to *interpose* safety barriers between the hazards and the potential losses
- to *contain* and *eliminate* the hazards should they escape this barrier
- to provide the means of *escape* and *rescue* should hazard containment fail.

Implicit in the ordering of this list is the idea of 'defences-in-depth'— successive layers of protection, one behind the other, each guarding against the possible breakdown of the one in front. When understanding, awareness and procedural guidance fail to keep potential victims away from hazards, alarms and warnings alert them to the imminent danger and direct the system controllers (or engineered safety features) to restore the system to a safe state. Should this not be achieved, physical barriers stand between potential losses and the hazards. Other defences act to contain and eliminate the hazards. Should all of these prior defences fail, then escape and rescue measures are brought into play.

It is this multiplicity of overlapping and mutually supporting defences that makes complex technological systems, such as nuclear power plants and modern commercial aircraft, largely proof against single failures, either human or technical. The presence of sophisticated defences-in-depth, more than any other factor, has changed the character of industrial accidents. In earlier technologies, there were—and to the extent that they continue to operate, still are—relatively large numbers of individual accidents. In modern technologies, such as nuclear power generation and air transportation, there are very few individual accidents. Their greatest danger comes from rare, but often disastrous, organizational accidents involving causal contributions from many different people distributed widely both throughout the system and over time.

The defensive functions outlined above are usually achieved through a mixture of 'hard' and 'soft' applications. 'Hard' defences include such technical devices as automated engineered safety features, physical barriers, alarms and annunciators, interlocks, keys, personal protective equipment, non-destructive testing, designed-in structural weaknesses (for example, fuse pins on aircraft engine pylons) and improved system design. 'Soft' defences, as the term implies, rely heavily upon a combination of paper and people: legislation, regulatory surveillance, rules and procedures, training, drills and briefings, administrative controls (for example, permit-to-work systems and shift handovers), licensing, certification, supervisory oversight and—most critically—front-line operators, particularly in highly automated control systems.

In earlier technologies, human activities were primarily productive: people made or did things that led directly to commercial profit. However, the widespread availability of cheap computing power has brought about a dramatic change in the nature of human involvement in modern technologies. These changes are seen most starkly in nuclear power plants and 'glass cockpit' commercial aircraft. Instead of being physically and directly involved in the business of production (and hence in immediate contact with the local hazards), power plant operators and pilots act as the planners, managers, maintainers and the supervisory controllers of largely automated systems.[5] A crucial part of this latter role involves the defensive function of restoring the system to a safe state in the event of an emergency.

Defences-in-depth are a mixed blessing. One of their more unfortunate consequences is that they make systems more complex, and hence more opaque, to the people who manage and operate them. Human controllers have, in many such systems, become increasingly remote, both physically and intellectually, from the productive systems which they nominally control. This allows the insidious build-up of latent conditions, to be discussed later.

The 'Swiss Cheese' Model of Defences

In an ideal world all the defensive layers would be intact, allowing no penetration by possible accident trajectories—as shown on the left-hand side of Figure 1.4. In the real world, however, each layer has weaknesses and gaps of the kind revealed on the right-hand side of the figure. The precise nature of these 'holes' will be discussed in the next section; here, it is necessary to convey something of the dynamic nature of these various defences-in-depth.

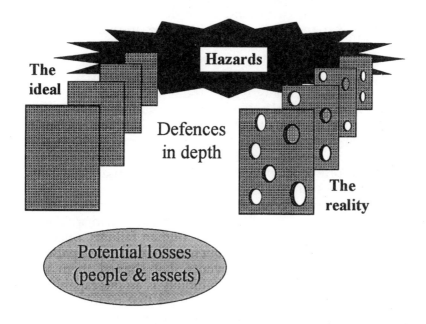

Figure 1.4   The ideal and the reality for defences-in-depth

Although Figure 1.4 shows the defensive layers and their associated 'holes' as being fixed and static, in reality they are in constant flux. The 'Swiss cheese' metaphor is best represented by a moving picture, with each defensive layer coming in and out of the frame according to local conditions. Particular defences can be removed deliberately during calibration, maintenance and testing, or as the result of errors and violations. Similarly, the holes within each layer could be seen as shifting around, coming and going, shrinking and expanding in response to operator actions and local demands.

How are the 'holes' created? To answer this, we need to consider the distinction between active failures and latent conditions.[6]

Active Failures and Latent Conditions

Since people design, manufacture, operate, maintain and manage complex technological systems, it is hardly surprising that human decisions and actions are implicated in all organizational accidents. Human beings contribute to the breakdown of such systems in two ways. Most obviously, it is by errors and violations committed at the 'sharp end' of the system—by pilots, air traffic controllers, police officers, insurance brokers, financial traders, ships' crews, control room operators, maintenance personnel and the like. Such unsafe acts are likely to have a direct impact on the safety of the system and, because of the immediacy of their adverse effects, these acts are termed *active failures*.

If these were individual accidents, the discovery of unsafe acts immediately prior to the bad outcome would probably be the end of the story. Indeed, it is only within the last 20 years or so that the identification of proximal active failures would not have closed the book on the investigation of a major accident. Limiting responsibility to erring front-line individuals suited both the investigators and the organizations concerned—to say nothing of the lawyers who continue to have problems with establishing the causal links between top-level decisions and specific events.

Today, neither investigators nor responsible organizations are likely to end their search for the causes of an organizational accident with the mere identification of 'sharp-end' human failures. Such unsafe acts are now seen more as consequences than as principal causes.[7] These developments and the events that shaped them will be discussed in Chapter 4. Although fallibility is an inescapable part of the human condition, it is now recognized that people working in complex systems make errors or violate procedures for reasons that generally go beyond the scope of individual psychology. These reasons are *latent conditions*.

Latent conditions are to technological organizations what resident pathogens are to the human body. Like pathogens, latent conditions—such as poor design, gaps in supervision, undetected manufacturing defects or maintenance failures, unworkable procedures, clumsy automation, shortfalls in training, less than adequate tools and equipment—may be present for many years before they combine with local circumstances and active failures to penetrate the system's many layers of defences. They arise from strategic and other top-level decisions made by governments, regulators, manufacturers, designers and organizational managers. The impact of these decisions spreads throughout the organization, shaping a distinctive corporate culture (see Chapter 9) and creating error-producing factors within the individual workplaces.

Latent conditions are present in all systems. They are an inevitable part of organizational life. Nor are they necessarily the products of bad decisions, although they may well be. Resources, for example, are rarely distributed equally between an organization's various departments. The original decision on how to allocate them may have been based on sound commercial arguments, but all such inequities create quality, reliability or safety problems for someone somewhere in the system at some later point. No single group of senior managers can foresee all the future ramifications of their current decisions.

A very important distinction between active failures and latent conditions thus rests on two largely organizational factors. The first has to do with the time taken to have an adverse impact. Active failures usually have immediate and relatively shortlived effects whereas latent conditions can lie dormant for a time doing no particular harm until they interact with local circumstances to defeat the system's defences. The second difference between them relates to the location within the organization of their human instigators. Active failures are committed by those at the human–system interface—the front-line or 'sharp-end' personnel. Latent conditions, on the other hand, are spawned in the upper echelons of the organization and within related manufacturing, contracting, regulatory and governmental agencies.

Whereas particular active failures tend to be unique to a specific event, the same latent conditions—if undiscovered and uncorrected—can contribute to a number of different accidents. Latent conditions can increase the likelihood of active failures through the creation of local factors promoting errors and violations. They can also aggravate the consequences of unsafe acts by their effects upon the system's defences, barriers and safeguards.

The Accident Trajectory

The necessary condition for an organizational accident is the rare conjunction of a set of holes in successive defences, allowing hazards to come into damaging contact with people and assets. These 'windows of opportunity' are rare because of the multiplicity of defences and the mobility of the holes. Such an accident trajectory is shown in Figure 1.5. Active failures can create gaps in the defences in at least two ways. First, front-line personnel may deliberately disable certain defences in order to achieve local operational objectives. The most tragic instance of this was the decision by the control room operators to remove successive layers of defence from the Chernobyl RBMK nuclear reactor in order to complete their task of testing a new voltage generator. Second, front-line operators may unwittingly fail in

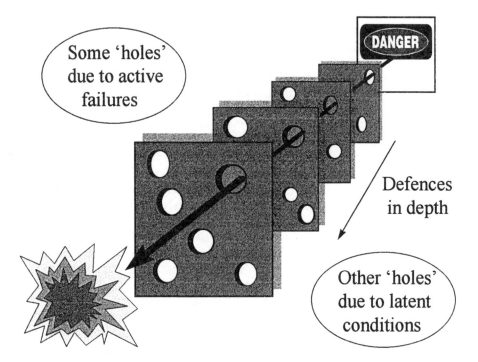

Figure 1.5   An accident trajectory passing through corresponding holes in the layers of defences, barriers and safeguards.
The holes can be created by active and latent failures.

their role as one of the system's most important lines of defence. A common example would be the wrong diagnosis of an off-normal system state, leading to an inappropriate course of recovery actions. Such 'sharp-end' mistakes played a significant role in the Three Mile Island nuclear power plant accident, the Bhopal methocyanate disaster, the Heysel and Sheffield football stadium crushes[8] and numerous other organizational accidents.

Since no one can foresee all the possible scenarios of disaster, it is therefore inevitable that some defensive weaknesses will be present from the very beginnings of a system's productive life, or will develop unnoticed—or at least uncorrected—during its subsequent operations. Such latent conditions can take a variety of forms: examples are the capsize-prone design of the *Herald of Free Enterprise*, the defective O-rings in *Challenger's* booster rockets, the inadequate fire containment and corroded sprinklers on the *Piper Alpha* gas platform, the failure to appreciate the risk of fire in London Underground stations, and the gradual erosion of supervisory checks that led both to the Clapham Junction rail disaster and the collapse of Baring's Bank.

Chapter 2 will present three case studies that illustrate some of the different ways that active failures and latent conditions can bring about the partial or total penetration of a system's defences. In the meantime, we need to consider the stages involved in the history of an organizational accident.

From the Deadwood Stage to Jumbo Jet

We can think about the history of organizational accidents over two different timespans: the one covering the history of a particular organization, and the other covering the history of a particular organizational accident. Although our main interest is in the latter, it is worth giving some attention to the former, if only to illustrate further the distinction between individual and organizational accidents.

Let us consider an imaginary organization that began its life more than 100 years ago in the American West, delivering passengers and mail from one frontier town to another by stagecoach. Let us also assume that the same company is today successfully operating a worldwide air freight business, using contemporary cargo jets and all the hi-tech paraphernalia that now accompanies such a venture. (Any resemblance to an actual organization is entirely coincidental.)

In the beginning, the founding fathers would have raised the money to buy a stagecoach and a few teams of horses. They would probably have done the driving themselves—at least at the outset. Their business would expose them to many hazards: hostile Native Americans, bandits, deserts, inadequate tracks, ravines and a variety of extreme weather conditions. Some of these dangers they could anticipate; others they would discover through bitter experience. Over time, their protection would gradually increase—from issuing the drivers with a hand gun and a sheepskin coat, to hiring additional people to ride shotgun and providing them with better weapons, to getting cavalry escorts through dangerous territory, and so on.

If disaster struck in these pioneering days, the fault—if such it could be called—lay almost entirely with the people on the spot. A driver could ignore local warnings and select a dangerous route, or overturn his coach by taking a bend too fast; the guard's gritty Winchester rifle could jam, or run out of ammunition, or he could simply fail to shoot straight.

At this stage in the organization's history, all such mishaps would be individual accidents, having very localized causes and consequences. Now consider a contemporary scenario. Suppose one of the company's jumbo jets takes off from a large city airport and then crashes into an apartment building, killing both the crew and many residents. The first possibility raised after such an accident is that of 'pilot error'. In this case, however, the air accident investigators soon establish that a fuse pin in one of the engine pylons failed just after takeoff, causing the engine to fall off, rendering the plane uncontrollable. Subsequent checks in the company's maintenance facility reveal that the aircraft had just undergone a major overhaul entailing the non-destructive testing of the engine fuse pins. It is also found that the fuse pin retainers on this particular engine had not been replaced following the service and their absence had not been spotted by the inspector. Access to the fuse pin area of the engine pylon was reached by an unstable underwing work platform and the area was poorly illuminated. The disassembly and replacement of the fastenings was the responsibility of a technician who was poorly trained and did not appreciate the need for red tag warnings to show the non-routine removal of parts—and so on.

The investigators might then go on to ask why the technicians received no formal classroom training, why the Director of Training's post was currently vacant, why the workplace was inadequate, and why the manufacturers of the aircraft felt it necessary to install fuse pins in the first place since, although they were intended as a safety device (allowing the ready separation of the engine from the wing in the event of an explosion or collision), their failure has been implicated in many incidents and accidents.

In short, this was an event for which no one person's failure was a sufficient cause. It was an organizational accident whose origins could be traced far back into many parts of the air cargo system, from the operator to the manufacturer, and—by implication—to the regulator. Should anyone think that such a combination of unhappy events is too unlikely to be credible, it must be pointed out that all of the individual failures catalogued in this story have actually happened, though not in a single aircraft accident.

The message of this hypothetical case study is that, while it may have been appropriate for those concerned with individual accidents in low-tech systems to focus upon the unsafe acts of those involved, such an approach would entirely fail to find the remediable causes of an organizational accident. In a stagecoach mishap, virtually all the causes would be active failures. Finding the means to correct those failings—better weapons, better suspension, better maps, better weather protection—and the safety margins increase, or would do if these defensive improvements did not encourage drivers to take short-cuts through territory hitherto regarded as too dangerous (and that is a very big 'if').

It should be apparent by now that a similar person-centred investigation in the case of the jumbo jet crash would have little or no chance of improving the system's safety. So long as people continue to be employed in modern technological systems, there will always be active failures—but very few of them will have bad consequences because most will be caught by the defences. We cannot change the human condition, but we can change the conditions under which people work. However, radical improvements of this kind can only be achieved through a better understanding of the nature of organizational accidents. Some preliminary steps towards this goal are presented below.

Stages in the Development of an Organizational Accident

The story of the cargo jet crash, just presented, gives some clue as to the difficulties facing those who seek to track down the root causes of an organizational accident. Where do you draw the line? At the organizational boundaries? At the manufacturer? At the regulator? With the societal factors that shaped these various contributions? As we shall see later, all of these increasingly remote influences are being considered by contemporary accident inquiries. In theory, one could trace the various causal chains back to the Big Bang. What are the stop rules for the analysis of organizational accidents?

Since time and causality are seamless, they have no natural breakpoints, only artificially imposed ones. Accident analysts, just like historians, are limited by their resources and by the availability of reliable evidence. Leaving aside legal concerns with responsibility, accident investigations are carried out for two main reasons: to establish what occurred and to stop something like it happening in the future. Both of these ends are best satisfied by limiting the scope of the analysis to those things over which the people involved—and most particularly the system managers—might reasonably be expected to exercise some control. A sad little story will help to make this point clearer.

Academician Valeri Legasov was the principal investigator of the Chernobyl reactor disaster that occurred on 26 April 1986. He was also the Soviet Union's chief spokesman at the international conference on this accident, held in Vienna in September of the same year. At that meeting Legasov put the blame for the disaster squarely on the errors and violations of the operators. Later, he confided to friends, 'I told the truth in Vienna, but not the whole truth.' In April 1988, two years to the day after the disaster, he hanged himself from the balustrade of his apartment. Prior to his suicide, he confided his innermost feelings about the accident to a tape recorder.

> After being at Chernobyl, I drew the unequivocal conclusion that the accident was ... the summit of all the incorrect running of the economy which had been going on in our country for many years.[9]

Legasov may well have been right in his evaluation. But how does this information help us? We can hardly go back to 1917 and replay the years following the Russian Revolution. Although Legasov's verdict was correct, it was unlikely to lead to achievable improvements. Models of accident causation can only be judged by the extent to which their applications enhance system safety. The economic and societal shortcomings, identified by Legasov, are beyond the reach of system managers. From their perspective such problems are given and immutable. But our main interest must be in the changeable and the controllable.

For these reasons, and because the quantity and the reliability of the relevant information will deteriorate rapidly with increasing distance from the event itself, the accident causation model presented in Figure 1.6 must, of necessity, be confined largely to the manageable boundaries of the organization concerned. It should also be appreciated, however, that all technological organizations have close ties with other organizations. The wider aviation system, for example, includes carriers, airport authorities, maintainers, manufacturers, air traffic controllers, regulators, unions, civil service agencies, government ministers, and international bodies such as the International Air Transport Association and the International Civil Aviation Organization—all of whom may have some part to play in an organizational accident.

The principal stages involved in the development of an organizational accident are shown in Figure 1.6. This model seeks to link the various contributing elements into a coherent sequence that runs bottom-up in causation, and top-down in investigation.

Accepting the time-frame discussed earlier, the causal story starts with the organizational factors: strategic decisions, generic organizational processes—forecasting, budgeting, allocating resources, planning, scheduling, communicating, managing, auditing, and the like. These processes will be coloured and shaped by the corporate culture, or the unspoken attitudes and unwritten rules concerning the way an organization carries out its business (see Chapter 9 for a further discussion of organizational culture).

The consequences of these activities are then communicated throughout the organization to individual workplaces—control rooms, flight decks, air traffic control centres, maintenance facilities and so on—where they reveal themselves as factors likely to promote unsafe acts. These include undue time pressure, inadequate tools and equipment, poor human–machine interfaces, insufficient training, under-manning, poor supervisor–worker ratios, low pay, low status,

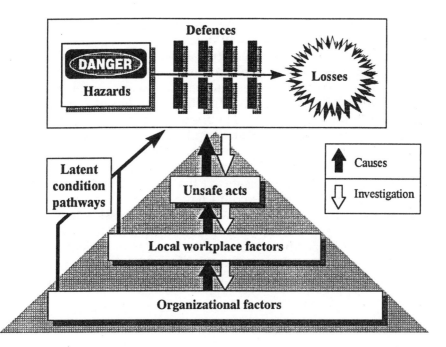

**Figure 1.6   Stages in the development and investigation of an
organizational accident**
The rectangular block at the top represents the main
elements of an event while the triangular shape below
represents the system producing it. This has three levels:
the person (unsafe acts), the workplace (error-provoking
conditions), and the organization. The black upward arrows indicate the direction of causality and the white
downward arrows indicate the investigative steps.

macho culture, unworkable or ambiguous procedures, poor communications and the like.

Within the workplace, these local factors combine with natural
human tendencies to produce errors and violations—collectively
termed 'unsafe acts'—committed by individuals and teams at the
'sharp end', or the direct human–system interface. Large numbers of
these unsafe acts will be made, but only very few of them will create
holes in the defences. The nature of these unsafe acts will be considered in detail in Chapter 4. Although unsafe acts are implicated in
most organizational accidents, they are not a necessary condition. On

some occasions, the defences fail simply as the result of latent conditions—as, for example, in the *Challenger* and King's Cross Underground fire disasters. This possibility is indicated in Figure 1.6 by the latent condition pathways, connecting workplace and organizational factors directly to failed defences.

So far, we have considered the causal sequence, from organizational factors, to local workplace conditions, to individual (or team) unsafe acts, to failed defences and bad outcomes. In the analysis or investigation of accidents, the direction is reversed (i.e., along the white arrows in Figure 1.6). The inquiry begins with the bad outcome (what happened) and then considers how and when the defences failed. For each breached or bypassed defence, it is necessary to establish what active failures and latent conditions were involved. And for each individual unsafe act that is identified, we must consider what local conditions could have shaped or provoked it. For each of these local conditions, we then go on to ask what upstream organizational factors could have contributed to it. Any one local condition could be the product of a number of different organizational factors, since there is likely to be a many-to-many mapping between the organizational and workplace elements of the model.

Pulling the Threads Together

This chapter began by distinguishing between individual and organizational accidents. Although the frequency of individual accidents in the workplace has decreased dramatically over the years, their basic nature—unprotected slips, lapses, trips and fumbles—has remained more or less unchanged in that the individual (or work group) is likely to be both the agent and victim of the accident. By contrast, organizational accidents—the topic of this book—are a relatively new phenomenon, stemming from technological developments that have both greatly increased the system's defences and changed the role of people from distributed makers and doers to centralized thinkers and controllers.

In contrast to earlier times, the human elements of modern technologies are often distanced from the day-to-day hazards of their respective operations. Organizational accidents in hi-tech systems occur very rarely, but when they do happen, the outcomes are likely to be disastrous, affecting not only those immediately involved but also people and assets that are far removed from the event in both time and distance. The Chernobyl fallout, for example, continues to remain a threat to unborn generations over large areas of Europe.

Understanding and limiting the occurrence of organizational accidents are the major challenges to be faced in the new millennium.

They are unacceptable in terms of their human, environmental and commercial costs. But how do we develop a set of concepts that are equally applicable to all of these highly individual and infrequent events, and—most importantly—lead to improved prevention? Two sets of terms have been proposed. The first provides a framework for understanding the details of an individual event—hazards, defences and losses. The second—the tension between production and protection—offers a means of understanding the processes that lead to defensive failures. Two such processes were the trading off of protective gains for productive advantage and the gradual deterioration of defences during periods in which the absence of bad events creates the impression that the system is operating safely—the case of the 'unrocked boat'.

Defences-in-depth have made modern technological systems largely immune to isolated failures. As such, they are the single feature most responsible for the emergence of organizational accidents. We have categorized these barriers and safeguards in two ways: by their functions (awareness, understanding, warning, guidance, restoration, interposition, containment, escape and rescue) and by their modes of application ('hard' and 'soft' defences). However, no one defensive layer is ever entirely intact. Each one possesses gaps and holes created by combinations of active failures (errors and violations committed by front-line personnel) and latent conditions (the consequences of top-level decisions having a delayed-action effect upon the integrity of various defensive layers).

The chapter concluded with two perspectives on the development of organizational accidents. The first outlined the development of a hypothetical organization from its early pioneering days, when all mishaps were likely to be individual accidents, until modern times, when the greatest risk was from rare but catastrophic organizational accidents. The second perspective covered the history of a particular organizational accident and traced its development from upstream systemic factors, through local factors promoting errors and violations, to the level of the individual who committed these unsafe acts. Some of these active failures are likely to have an immediate adverse effect upon local defences, many others will be inconsequential. When the gaps produced by active failures 'line up' with those created by latent conditions, a 'window of opportunity' exists for an accident trajectory to bring hazards into damaging contact with people and assets.

The main purpose of this chapter has been to introduce the basic ideas and to set the stage for a more thorough consideration of these issues in subsequent chapters. Chapter 2 begins this finer-grained analysis by examining the various 'cracks in the system' revealed by case studies of three organizational accidents. Our purpose here is to put some flesh on these rather abstract bones. It must be stressed, however, that this is not yet another book of accident case studies. Their sole reason for inclusion here is to achieve a better understanding of organizational accidents in general rather than in the particular instance.

Notes

1   Individual accidents can, and usually do, have organizational origins. Indeed, as Andy Pearce of Shell Expro has pointed out, an offshore fatality is typically the result of a chain of at least seven distinct failures. Although it is not always easy to draw a hard and fast line between individual and organizational accidents, this book argues that it is useful to treat them as distinct kinds of event.
2   In simpler technologies, the productive and protective elements were often different structures. But in complex technologies, the same entity can serve both productive and defensive functions—as, for example, in the case of pilots and control room operators.
3   These and other examples of risk compensation are discussed by Erik Hollnagel in *Human Reliability Analysis: Context and Control*, (London: Academic Press, 1993, pp. 8–11). See also G.J.S. Wilde 'The theory of risk homeostasis: implications for safety and health.' *Risk Analysis*, 2, 1982, pp. 209–55. An excellent critique of the idea of risk homeostasis as it applies to road transport has been given by Leonard Evans in *Traffic Safety and the Driver*, (New York: Van Nostrand Reinhold, 1991, pp. 298–300).
4   The phrase was coined by Constance Perin in a paper entitled 'British Rail: the case of the unrocked boat'. This commentary on the Clapham Junction railway accident was given to a Workshop on Managing Technological Risk in Industrial Society, 14–16 May 1992, at Bad Homburg, Germany.
5   See N. Moray, 'Monitoring behaviour and supervisory control' in K. Boff, L. Kaufman and J. Thomas (eds), *Handbook of Perception and Human Performance*, vol. 2 (New York: Wiley, 1986).
6   See J. Reason, *Human Error*, (New York: Cambridge University Press, 1990), ch. 7. When the term first appeared it was as 'latent errors', then it changed to 'latent failures. Now the term 'latent conditions' is preferred, since it does not necessarily involve either error or failure.
7   The idea of errors as consequences rather than causes has been developed by David Woods and his colleagues at Ohio State University. See, for example, D. Woods *et al.*, *Behind Human Error: Cognitive Systems, Computers and Hindsight. State-of-the-Art Report*, (Dayton, Ohio: CSERIAC, Wright-Patterson Air Force Base, 1994).
8   While the other accidents listed here will be discussed elsewhere in the book (and sources provided), no further mention will be made of football stadia disasters. Perhaps the best analysis of these accidents has been given by Dominic Elliott and Denis Smith in their paper 'Football stadia disasters in the United Kingdom: learning from tragedy', *Industrial and Environmental Crisis Quarterly*, 7, 1993, pp. 205–29.
9   This quotation is taken from the *Legasov Tapes*, a transcript prepared by the US Department of Energy in 1988. The original material appeared in *Pravda* shortly after Legasov's death.

2 Defeating the Defences

Things are not Always What They Seem

The human mind is prone to match like with like[1]. It is therefore natural for us to believe that disastrous accidents must be due to equally monstrous blunders. But the close investigation of organizational catastrophes has a way of turning conventional wisdom on its head. Defences can be dangerous. The best people can make the worst mistakes. The greatest calamities can happen to conscientious and well run organizations. Most accident sequences, like the road to hell, are paved with good intentions—or with what seemed like good ideas at the time.

In the previous chapter, we introduced the gaps created in the system's defences by active failures and latent conditions. When they lead to a particularly bad outcome, we naturally assume that it could only have been the result of gaping holes due to gross negligence and habitual bad practice. Bad practices may be present, but the reality is often much more commonplace. Organizational accident trajectories can just as easily slip through the small and apparently insignificant cracks in the system as they can through the yawning gaps.

This chapter examines three recent case studies drawn from widely differing domains. The first near-accident slipped through the minor cracks in the system; the second disaster fell through some yawning gaps in the defences; the third, unusually, had its origins in a single longstanding latent condition. Many other well documented case studies would have served as well, but these—for the most part—have already been extensively analysed elsewhere.

Our purpose is to try to identify recurring patterns in the way defences fail. Are there sufficient common features, even in these disparate events, for us to abstract some general principles regarding the development of organizational accidents? The American social scientist, Karl Weick, stated the nub of the problem very well:

We know that single causes are rare, but we don't know how small events can become chained together so that they result in a disastrous outcome. In the absence of this understanding, people must wait until some crisis actually occurs before they can diagnose a problem, rather than be in a position to detect a potential problem before it emerges. To anticipate and forestall disasters is to understand regularities in the ways small events can combine to have disproportionately large effects.[2]

Slipping Through the Cracks of an Aircraft Maintenance System

Our first case study is a prime example of an aircraft accident that nearly found its way through the small chinks in the system's extensive armour. Fortunately, there was no actual catastrophe, but it was a close-run thing, only avoided by the prompt action of the last-ditch human defences.[3]

Shortly after departing from East Midlands Airport en route for Lanzarote in the Canary Islands, the pilots of a Boeing 737-400 detected the loss of oil quantity and oil pressure on both engines. They declared a 'Mayday' and diverted to Luton Airport, where both engines were shut down during the landing roll. There were no casualties. It was later discovered that the high-pressure rotor drive covers on both engines were missing, resulting in the almost total loss of the oil from both engines during flight. A scheduled borescope inspection (required every 750 hours) had been carried out on both engines during the previous night by the airline's maintenance engineers.

The borescope inspections were to be performed by the line maintenance night shift. On the night before this work was to be done, the line engineer in charge of this shift had expressed his concern about the manpower available to carry out the next night's predicted workload, which he knew would include the borescope inspections. However, on arriving for work on the night in question, he discovered that no extra maintenance personnel had been assigned to his shift. Instead of a nominal complement of six, there were only four on duty that night and two of them—including the line engineer— were working extra nights to cover shortfalls.

The line engineer realized that he would have to carry out the borescope inspections himself since he was the only engineer on the line shift possessing the necessary authorization. As the inspection was to be carried out in a hangar at some distance from where the bulk of the line maintenance work was done, but close to the base maintenance hangar, he decided to put the inspection at the top of his list of jobs for that night.

After organizing the night's work for his shift, he collected the inspection paperwork from the line office and went to the aircraft where he started to prepare one of the engines for inspection. Having done this, he went across to the nearby base maintenance hangar where the borescope equipment was stored. There he met the base maintenance controller (in charge of the base night shift) and asked for the inspection equipment and also for someone to help him, since the engine spool had to be turned by a second person as he carried out the inspection.

The base night shift was also short-staffed. On arriving at his office that night, the base controller had received a request from Ramp Services to remove another aircraft from their area (a Boeing 737-500) because it was in the way. He could not do this immediately because of the shortage of personnel on his shift. However, when he met the line engineer, he saw a way of killing two birds with one stone. The company rules required staff to carry out two 750-hour borescope inspections[4] within a 12-month period in order to maintain their borescope authorization. His own authorization was in danger of lapsing because this task very rarely came his way. He also knew that the line maintenance shift was lacking two people and that they had eight aircraft to deal with that night. So he offered to do a swap. He would carry out the borescope inspection, if the line engineer took over the job of moving the B737-500 from the ramp to base.

The line engineer agreed to this mutually convenient arrangement and told the base controller of his progress so far in the preparation of the engines. Since the line engineer had not visited the base hangar with any intention of handing over the inspection to someone else, there was no written statement or note of what had so far been done. Indeed, the line maintenance paperwork offered no suitable place to record these details. The base controller, however, was satisfied with the verbal briefing that he had received. His next step was to select a fitter to assist him and to check and prepare the borescope equipment.

With these innocent and well intentioned preliminaries, the stage was set for a near-disaster. The fitter was sent to prepare the second engine for inspection and the base controller then spent a considerable time organizing the duties for his short-staffed night shift. When he eventually joined the fitter at the aircraft, he brought with him his personal copy of the borescope training manual on which he had written various details to help him in judging sizes through the probe. Although the work pack for the job was unfamiliar to him— being line rather than base maintenance paperwork—and although there were no Boeing task cards attached, he did not consider it necessary to draw any additional reference material. While the fitter was continuing to prepare the second engine for inspection, the base controller started his inspection of the first engine.

Throughout the next few hours, the job continued with many interruptions brought on by the base controller's need to cope with pressing problems cropping up elsewhere on his patch. The upshot was that the rotor drive covers were not refitted, the ground idle engine runs (checks that could have revealed the oil leaks) were not carried out, and the borescope inspections on both engines were signed off as being complete in the Aircraft's Technical Log.

During the night, the line engineer returned with a line maintenance crew to prepare the aircraft for the next day's flying. He and the base controller discussed whether or not to trip out the aircraft's engine ignition and hydraulic circuit breakers. The base controller said that he did not think it was necessary since he had no intention of working on the aircraft with the electrical and hydraulic power systems active, but the line engineer pulled the circuit breakers anyway, feeling confident that the base controller was aware of this. When the aircraft was later returned to the line, the pilots noted that these circuit breakers had not been reset and expressed some dissatisfaction about this to a line engineer on the morning shift. The engineer said that it was probably an oversight and reset them. When the engineer returned to the line office, he wondered aloud to a colleague how the engine run had been done with the ignition circuits disabled. They were still discussing this when they heard that the aircraft had made an emergency landing at Luton.

We do not know exactly why the covers were not replaced on the engines, nor how the inspection came to be signed off as completed when such vital parts were missing and the engine run not done. We do know, however, that omissions during reassembly are the single most common form of maintenance lapse—both in aviation and in the nuclear power industry. Indeed, such maintenance omissions are probably the largest single type of human performance problem in most hazardous technologies (see Chapter 5). We know too that there had been at least nine previous instances in which high-power rotor covers had been left off engines following maintenance in other airlines. The available evidence also indicates that many aircraft maintenance jobs are signed off as complete without a thorough inspection. Such routine short-cuts are most likely to occur in the face of a pressing deadline—such as returning an aircraft to the line before its scheduled departure time.

Knowing precisely what went on in the minds of the people concerned at the time these errors were committed would contribute little to our understanding of how organizational accidents happen or how their occurrence could be thwarted in the future. Such human failures happen frequently, and will continue to do so while people are still engaged in removing and replacing the millions of detachable parts that make up a modern aircraft. Our primary concern, both here and in other organizational accidents, is with the systemic factors that promoted their occurrence and then allowed them to go undetected. Human fallibility, like gravity, weather and terrain, is just another foreseeable hazard in aviation. The issue is not why an error occurred, but how it failed to be corrected. To repeat a constant theme of this book: We cannot change the human condition, but we can change the conditions under which people work.

What, then, were the small cracks in the system that permitted these commonplace deviations to come so close to causing a major air disaster? The first and most obvious factor is that both the line and base maintenance crews were short-staffed, though not so depleted that they had to abandon their assigned tasks. Such temporary shortfalls in maintenance personnel through sickness and leave are not uncommon. A similar problem in a separate incident contributed to the blow-out of a flight deck windscreen that had been secured by the wrong-sized bolts[5]. There, as in this case, a temporary shortage of personnel led supervisory staff to take on jobs for which they lacked the current skills and experience. And, in both events, their ability to perform these tasks reliably was impaired by the continuing need to carry out their managerial duties. To compound these problems further, the shift managers involved in both accidents were also in a position to sign off their own incomplete work, thus removing an important opportunity for detecting the error.

In the case of the missing engine covers, however, there was also another more subtle crack in the system—the difference in working practices between line and base maintenance. The division of labour between these two arms of an aircraft maintenance facility is normally clear-cut. Line engineers are accustomed to performing isolated and often unscheduled tasks (defect repairs) either individually or in small groups. These jobs are usually completed within a single shift. Base maintenance, on the other hand, deals with scheduled overhauls. When an aircraft enters a base maintenance hangar, it is likely to be there for a relatively long time and a large number of different tasks will be carried out by rotating shifts, with incomplete work being handed on from one shift to the next.

The line and base environments thus require different kinds of planning and supportive work packs. In the case of line maintenance, the paperwork is generated just before the work is due to be done and is subject to change according to operational needs. The job cards therefore give only a brief description of the work to be performed. In sharp contrast, planning for base maintenance work starts several weeks before the event and is delivered as a massive work pack that includes considerable supporting and explanatory documentation. Life on the line is full of surprises, while that in base maintenance is far more regular and predictable.

The 750-hour borescope inspection was designated as a line maintenance task, and the paperwork included none of the step-by-step task cards and maintenance manual sections normally supplied with a base maintenance work pack. It only contained references to the detailed documentation that was available elsewhere. The line engineer was familiar with the job and did not feel it necessary to draw this additional information, so the work pack that he handed over to the base controller was relatively sparse in comparison with the usual base maintenance documentation. In particular, the line-generated work pack contained no mention of the restorative work —replacing fasteners and covers—nor did it require step-by-step signatures confirming the completion of these tasks, as would be expected in a base-generated work pack. Nor were there any of the customary warnings—highlighted in the maintenance manual—that the safety of the aircraft would be seriously jeopardized if the reassembly work was not completed as specified.

Given the short-staffed nature of his own shift and the additional demands that this would inevitably make upon his time, the base controller was probably unwise to offer to carry out the borescope inspection. But it could hardly be classed as a major blunder, and it was certainly allowable within company procedures. In view of his lack of current experience at the task, he was clearly mistaken in assuming that he could perform the task without reference to detailed task cards and using only his memory and the unofficial training notes as a guide. Until he actually received the work pack from the line engineer, he probably would not have known that the detailed supporting material was missing. At this point, he could have gone across to the line maintenance office or to base maintenance to fetch it, but it was a cold winter's night and he was confident that he knew the procedure. And even the accident investigator acknowledged that had he retrieved this material, it would not have been easy to use. All manuals are continually being amended and updated, often making it hard for those unfamiliar with their layout to follow a continuous thread of task-descriptive text.

Another small crack in the system was the absence of any formal procedure for handing over jobs from line to base maintenance. As a result, no preprepared stage sheets (indicating the prior work performed) were available and the handover briefing was carried out verbally rather than in writing. This was not so much a case of a flawed defence as one that was entirely absent. Yet such handovers are comparatively rare events and, while it is easy to appreciate the need for handover procedures with hindsight, it is difficult to imagine why their creation should be high on anyone's agenda before this event occurred. These various cracks in the defences are summarized in Table 2.1.

Table 2.1   Summary of the active failures and latent conditions that undermined or breached the aircraft maintenance system's defences

Active Failures	Latent Conditions
The line maintenance engineer who released the aircraft to the flight crew missed an opportunity to discover the missing drive covers when he reset the ignition circuit breakers and queried how an engine run could have been performed.	The airline lacked an effective way of monitoring and adjusting the available manpower to the workload, particularly on night shifts known to be subject to adverse time-of-day effects.
The flight crew accepted the pulled circuit breakers on the flight deck as 'normal' and did not pursue the matter further.	The regulatory surveillance of this manpower problem was inadequate.
The aircraft's Technical Log was signed off without adequate inspection of the work on the engines.	Both the line and the maintenance night shifts were short-staffed. This was a fairly common occurrence.
The base controller was in a position to sign off his own incomplete work without further inspection.	An internal inquiry revealed that borescope inspections were often carried out in a non-procedural manner. This was a failure of the company's Quality Assurance system.
The base controller failed to supervise the work of the relatively inexperienced fitter sufficiently closely. This was partly due to the many interruptions arising from his administrative duties.	No procedures existed for the transfer of part-completed work from line to base maintenance. As a result, no stage paperwork was available, and the line engineer only gave a verbal handover briefing to the base controller.
No idle engine run was performed.	The line-oriented paperwork given to the base controller lacked the usual (for him) reference documentation. In particular, it lacked any reference to restorative work and provided no means of signing off such work.
The base controller carried out the work without reference to task cards or the maintenance manual. Instead, he relied on his memory and an unapproved reference—his training notes.	

In July 1996 the airline involved in this accident was fined £150 000 plus £25 000 costs for 'recklessly or negligently' endangering an aircraft and its passengers.[6] This was the first time in the UK that the Civil Aviation Authority had prosecuted a carrier under Articles 50 and 51 of the 1989 Air Navigation Order. The two maintenance engineers directly involved in the accident were dismissed. After the verdict, the deputy chairman of the airline told the press: 'This has been a difficult day for us, but we have learnt from the experience. We have completely changed our procedures.' We will consider whether or not this was the most appropriate response in Chapter 3.

The captain of the aircraft, whose prompt actions averted a catastrophe, stated that it was a pilot's job to cope with the unexpected: 'We have to anticipate the worst case scenario. We are not just up there to press a button and trust in the wonders of modern technology. We have to be ready for this kind of eventuality.'[7] This states the defensive role of flight crew very clearly. In Chapter 5 we pose the question: Is the primary task of pilots in contemporary aircraft to cope with emergencies produced by undiscovered maintenance errors?

The Millions that Gushed Away: the Barings Collapse

To those who work in mining, construction, transportation, nuclear power generation, chemical process plants and the like, the term 'hazard' usually means some inanimate danger associated with the uncontrolled release of mass, energy or toxic substances. But it is worth reminding ourselves that 'defences-in-depth' is a military term, relating to situations in which the hazards to be guarded against are other people.

People constitute a hazard in many ways. For the military, it is 'the enemy'. In a football crowd, it is the crush of other bodies. For commercial aviation, it takes the form of terrorists and hijackers. In big cities, it is muggers, thieves and murderers. For banks, it can be rogue traders.

The second case study concerns the collapse of the Barings banking group—or at least that part of the story relating to the failed defences. In keeping with the approach adopted earlier, we will not delve into the competence, motives or the morals of the principal hazard in this case study—Nick Leeson, the self-styled rogue trader.[8] We will take it as a given that anyone trading in millions on a daily basis is exposed to a variety of temptations and is capable of making some very costly mistakes. Our interest, as before, is in the manner in which the banking system's various barriers, safeguards and defences were absent, breached or bypassed. Inevitably, this account will be much abridged[9] and will limit itself to the period of the actual collapse, although its origins may probably be traced back for 100 years or so.

On 26 February 1995, the High Court in London appointed joint administrators to manage the affairs of Barings plc, the Barings Group parent company. This followed the discovery of the loss of £869 million by Barings Futures (Singapore) Pte Limited (BFS) incurred on the Singaporean and Japanese exchanges. Barings Brothers & Co., Limited (BB & Co.) was the longest established merchant banking business in the City of London. Since the original foundation of the partnership in 1792, it had remained independent and privately controlled.

Barings Securities had been operating in Singapore since 1987, but it was only in 1992 that it acquired seats on the Singapore International Monetary Exchange (SIMEX) with the intention of trading in the rapidly growing futures-and-options market. The new company established to conduct this business was BFS. Leeson was appointed to head the settlement operations. Most unusually, he was also asked to be Barings Futures' floor manager on SIMEX. This breached one of the basic principles of the securities business which was to keep the settlements and trading activities strictly separate. The job of the trader was to make money. That of the settlements clerk was to ensure that no errors occurred in the accounting and to fix them when they did. Giving Leeson this dual role was subsequently described by the Singaporean investigators as an 'ill-judged decision'.[10]

Between July 1992 and February 1995, the Barings Group were carrying out a radical restructuring of their hitherto autonomous businesses, each with different cultures and operations. Conflict had arisen between the more conventional banking operations at BB & Co and its subsidiary, Barings Securities Limited (BSL). BSL had been formed from the acquisition in 1984 of a UK-based stockbroker with strong connections in Asia. BSL had flourished in the heady days of the late 1980s, but by 1992 it was no longer so healthy. Its chief executive, Christopher Heath (once Britain's highest-paid worker), was faced with falling markets and a large staff, hungry for fat bonuses.

The situation came to a head on Sunday, 26 September 1992, a day that Heath later described as 'the stabbath'.[11] Heath relinquished his post as CEO and Peter Norris was appointed as chief operating officer. In March 1993, Heath left Barings and Norris was made chief executive officer. At the end of 1993, BB & Co and BSL were merged to form the Investment Banking Group, later to become Baring's Investment Bank (BIB). BIB was subdivided into four groups. The two relevant to the collapse were the Equity Broking and Trading Group, headed by Norris, and the Banking Group, led by George Maclean. Maclean was later to tell the Bank of England investigators:

> I believe the seeds of this [collapse] were sown when we went into BSL to bring the two companies together and made the assumption that the quality controls that we [BB & Co] had could quickly get installed there [BSL]. As it turned out that appears not to be true.[12]

After passing the local exchange examinations, Leeson began trading on the floor of SIMEX. Soon he was appointed as General Manager and Head Trader of BSF, although there later seems to have been much confusion in London as to what this title entailed. Heath, who had originally appointed him, believed that he was just a clerk transmitting orders received by phone or fax to the trader on the SIMEX floor. Leeson, however, had other ideas. Being an execution clerk gave him access to the trading floor. In his book, *Rogue Trader*, he gives us a clear idea of the impression this made upon him.

> When I first stepped out on to the trading floor, I could smell and see the money. Throughout my time at Barings I had been inching closer and closer to it, and in Singapore I was suddenly there. I'd been working in various back offices for almost six years, pushing paper money around, sorting out other people's problems. Now, out on the trading floor, I could work with instant money—it was hanging in the air right in front of me, invisible but highly charged, just waiting to be earthed.[13]

The confusion surrounding Leeson's dual role as settlements clerk and trader lies at the heart of the Barings collapse. Just as a lighted cigarette fell through the cracks in the wooden escalator to cause the King's Cross Underground fire, so Leeson fell through the cracks in the Barings' matrix management structure. In this, each of the business ventures—banking, equity broking, trading and so on—formed the verticals of the matrix, while the various offices and regional structures scattered around the world formed the horizontals. The theory was that any one individual was connected both to local management and to the London office. The reality in this case was that Leeson's activities were not closely monitored by anyone. Leeson's immediate bosses in Singapore not only failed to grasp the extent of his job specification, they were also reluctant to supervise him. He was always regarded as someone else's responsibility. His reporting lines were either blurred or non-existent. This problem was not helped by the top management in Singapore being located on the twenty-fourth floor of the new Ocean Tower block, while Leeson worked on the fourteenth floor.

In July 1992 BFS opened an error account. This was a standard procedure for holding disputed trades until the disagreement is settled, usually within 24 hours. It was given the number 88888 (the five-eights account). Five days later, Leeson instructed an independent computer consultant in Singapore to change the software at BFS. He wanted the five-eights account to be excluded from daily trading, position and price reports to London, and only the daily margin file—relating to cash or securities deposited with the exchange as collateral and as a means of settling profit and loss—to be passed on. This meant that when the daily margin file arrived in London, the automatic sorting system would not recognize the account number. As a result, no information would be transferred into Barings' internal reporting system, called First Futures.

The information held in London was presented on two computer screens. One showed the margin balances on the SIMEX computer system and included the margin file from the five-eights account. The second showed the balances transferred to the First Futures system. This was the only screen regularly scrutinized by the London office, and it had no record of the five-eights account. London only became aware of this secret account three days before the collapse. It was through this hole that hundreds of millions of pounds vanished.

Leeson then embarked on a frenzy of unauthorized trading in futures and options. By December 1992, his losses amounted to £208 million. Throughout these and the succeeding months the London office regarded Leeson as a star performer. He was given a £130 000 bonus in 1993 and one of £450 000 the following year. Barings assumed that his declared profits came from arbitrage—trading on the price differences between SIMEX and the Japanese exchanges. Because the London office believed that these trades were fully matched—offsetting each other—there appeared to be no real risk to the bank. By February 1995 the accumulated losses amounted to £830 million.

Leeson funded his losses in three ways. First, he used money lent by Barings Securities subsidiaries in Tokyo and London in the belief that it was being traded on their own accounts. Second, money was lent by BSL in London for margin payments to the exchanges—funds to cover unrealized losses. But no adequate steps were taken to verify the accounts or to check them with the trading records of clients. Third, as the losses escalated in January and February 1995, Leeson created artificial trades on the SIMEX computer system to cut the level of margin payments required by the exchange. He also covered his tracks by submitting false reports to London inflating BFS's profits and making false trading transactions and accounting entries.

In July and August 1994, a BSL internal audit team visited Singapore to review the operations of Barings' offices in the region. The report, delivered in December 1994, expressed some concern at the lack of segregation between BFS's front (trading) and back (settlement) offices. The team member who audited the Singapore office wrote that while there was no evidence that Leeson was abusing his dual role, '... the potential for doing so needs examining'.[14] The report recommended that Leeson should no longer be solely responsible for supervising the back office, but this still left him with the powers necessary for his deceptions. One of the main reasons for the audit was to probe Leeson's large profits. The report concluded that the profits were due to Leeson's exceptional abilities and expressed the worry that should Leeson be poached by a competitor it would '... spread the erosion of BFS's profitability greatly'. The auditors, along with the London management, continued to be mesmerised by Leeson's declared profits.

By this time, however, the London office was aware that tens of millions of pounds remained unaccountable due to unreconciled trades—the failure to match what should have been in an account with what actually was there. Tony Hawes, the Group Treasurer of BIB, took some comfort from the audit report: 'We thought there couldn't have been anything too wrong, or they'd have spotted it.' Later, he was to say that there was no excuse for not making reconciliation the highest priority. 'But there always seemed to be something else more pressing'. This sentiment crops up frequently in the history of organizational accidents.

By early February 1995 there were rumours in the markets about dealings in Japan and about possible client problems. The London office even received a concerned phone call from the Bank for International Settlements in Basle. But this did not worry the Barings' management because they believed—wrongly—that their positions on the Japanese markets were covered by equal and opposite positions on SIMEX.

Nonetheless, the pace of discovery quickened through February 1995. On 6 February, Tony Hawes and a colleague, Tony Railton, visited Singapore in the hope of clarifying a number of issues. Gradually, they uncovered enough to cause them great anxiety. On 23 February, Leeson did not return to the office. On the same day, Hawes looked at a computer printout and noticed 'an account called an error account with goodness knows how many transactions on it, all of them seemingly standing at tremendous losses'.[15] The next day, a Friday, Peter Baring, Chairman of Barings plc, met the Deputy Governor of the Bank of England and informed him that Barings had been the victim of a massive fraud. Over the weekend, the Bank of England tried unsuccessfully to save Barings. On the Sunday evening, the administrators were called in. The 203 year-old merchant bank had collapsed.

So far, we have focused on the failure of the internal controls. But what of the failure of the Bank of England's regulatory role? The Bank of England's investigative report[16] of the collapse criticized the supervision provided by one of its senior managers in charge of merchant banks, stating that this manager made an error of judgement in 1993 when he gave Barings the informal concession of allowing its dealings on the Osaka Stock Exchange to exceed 25 per cent of its capital. As a result, the manager in question resigned, as did Peter Baring and Andrew Tuckey, the Chairman and Deputy Chairman of Barings plc. Nick Leeson is currently serving a six-and-a-half-year sentence in a Singaporean prison for fraud. The assets and liabilities of the Barings Group were purchased by ING, a large Dutch banking and insurance group, for a token £1—but the real cost was £660 million. In June 1996, the Singaporean *Business Times* reported that Singapore's largest lawsuit, involving more than S$2 billion, had been served on BFS's external auditors, Coopers & Lybrand Singapore and Deloitte & Touche, for giving an 'unqualified clearance on the Baring group's reporting package of BFS for the year ended 31 December 1994'. The fallout will no doubt continue.

By using the five-eights account both to conceal his losses and to declare false profits to the London office—Leeson created a gaping hole in the system's defences. Clearly, Leeson himself was the principal architect of the active failures. Some idea of the extent of these destructive trades can be gained from looking at the period immediately following the Kobe earthquake on 17 January 1995—the event that triggered his final downfall. Leeson gambled that Japanese stocks would rise rather than fall, and on 20 January, he built up a long position (one that appreciated in value if the market price increased) of 10 014 Nikkei 225 futures contracts. On Monday 23 January, the Nikkei 225—an index based on 225 Japanese stocks traded on the Tokyo Stock Market—fell by 1175 points, but Leeson continued to buy. By 27 January, his holding had increased to 27 158 contracts.

On 23 January alone, Leeson's Nikkei futures position lost £34 million, while his options portfolio had lost £69 million since the earthquake. Leeson's response was to use cross-trades (a means of transferring buy-and-sell orders through the exchange between two clients belonging to the same firm) with the five-eights account to declare large dividends for Barings. Between 23 and 27 January, when he claimed to have made £5 million from arbitrage, he had actually lost £47 million.

Fraud is extremely difficult to counter, and incompetent fraud of these proportions is even more difficult to predict or withstand. In April 1994 the international banking world was shaken by the news that a Wall Street trader had been discovered by his employer to have declared a wholly false profit of $350 million. The reaction of Barings' top management was to conduct a review of its own risk controls. They discovered that these controls needed radical improvement. The board was told that such a tougher risk management system was being developed, but nothing more was heard of this plan until January 1995 when it was too late to avert disaster. Risk controllers were appointed in Hong Kong and Tokyo during 1994, but the Singapore management said they were not necessary.

The Bank of England report identified a number of warning signs that should have alerted Barings management to BFS's unauthorized activities. The report concedes that no single one of these indicators would have served as a sufficient warning, but collectively they should have raised the alarm in both London and Singapore. These indicators included the following:

- The lack of segregation between the front and back offices identified by the internal audit carried out in July and August of 1994.
- The high level of funding required to finance BFS's trading activities.
- The unreconciled balance of funds transferred from London to Singapore for margins.
- The very high apparent profits relative to the low level of risk as perceived by Baring's management in London.
- The discovery of an apparent receivable of approximately £50 million from one of BFS's customers, Spear, Leeds & Kellogg, that had been faked by Leeson in December 1994.
- A letter sent by SIMEX to BFS on 11 January 1995 that included specific references to the five-eights account. This was not passed on to London at the time.
- And a further letter from SIMEX to BFS, dated 27 January 1995, seeking reassurance that BFS was able to fund its margin calls in the event of an adverse market move. This was communicated to London, but was not acted upon for the reasons given earlier.

The Nakina Derailment: A 76-Year-old Latent Failure

If there is one defensive feature that dominates all others in railway operations, it is the stability of the roadbed—the ground upon which the rails are laid. Should it fail, the alignment and the holding power of the rails are lost and any train passing over this region is likely to be derailed. This is what happened in July 1992 at a bend near Nakina Ontario.[17] What distinguishes this accident from other derailments is that the root cause dated back to 1916 when the track was first laid.

Just after coming round a bend at the regulation 35 mph, the crew of a Canadian National freight train saw that the rails immediately ahead of them were suspended in mid-air and that the roadbed beneath them was completely washed out. Unable to stop in time, the train derailed, toppling into an adjacent pond. Two crew members were killed and the train driver was seriously injured.

The subsequent investigation, carried out by the Canadian Transport Safety Board, established from the original plans that the roadbed had been located on the corner of an existing beaver pond. The rails had been laid across a beaver dam—an unstable mix of silt and peat. Although the upper portions of the roadbed had been maintained and upgraded over the ensuing decades, the basic instability of their foundations had not been rectified.

Organizational accidents involve the interaction of latent conditions with local triggering factors. Two such local events precipitated the accident. First, the railway company implemented a policy of killing beavers in the vicinity of rail tracks—to minimize the problems of flooding and washouts of railway infrastructure associated with beaver activity. This meant that the longstanding dam was no longer maintained by the beavers. Second, it had been an unusually wet summer and water had built up behind the dam supersaturating the nearby roadbed. The unattended beaver dam, gradually weakened under the increasing pressure of the accumulated water, collapsed when a critical mass of the subgrade sludge was washed into the bed of the pond.

The investigators built a computer model of the beaver dam and tested its stability under various scenarios. This exercise established that a temporary loading on the dam created by a freight train passing over it would not have been sufficient to cause the collapse. The catastrophic destabilization was due entirely to a rapid two-metre fall in the water level on one side of the railway embankment. There were no visual indications that the roadbed had been undermined, nor could this have been predicted without a detailed geotechnical analysis. Since the track appeared to be stable to those carrying out routine inspections of its exterior features, no such analysis was performed nor were measures taken to strengthen the underlying bed.

Unusually, therefore, no active failures (errors and violations) on the part of either the train's crew or the track's maintainers were implicated in this accident. The root cause was a latent construction failure—permitted by the standards existing in 1916—that had lain dormant for 76 years until a set of local triggering conditions caused its intrinsic weaknesses to be fatally revealed.

Common Features

The purpose of presenting these three case studies was twofold. First, to demonstrate in detail the variety of ways that a system's defences can be degraded and defeated to cause organizational accidents. Second, to see what—if any—common principles could legitimately be applied to these three very diverse events. It is the latter issue that will occupy us in this concluding section. In this regard, it is worth remembering that, in safety science, the test of any general principle is a very practical one. Does it identify workable remedial applications? Theoretical abstractions are of little interest unless they lead to improved safety.

The first and most obvious fact about all three of these accidents is that pre-existing—and often longstanding—latent conditions contributed to the breakdown of the system's defences. Errors and violations committed by those at the sharp end are common enough in organizational accidents, but they are neither necessary nor sufficient causes, as the Nakina derailment clearly showed. Latent conditions, however, are *always* present in complex systems. And they are present *now*. We may not know which particular conditions will be implicated in any future event, but we do know that some latent conditions will be involved. From the now extensive analyses of organizational accidents, we also have a fairly good idea of the *kinds* of latent conditions that are most likely to constitute a threat to the safety of the system. There is no mystery to this. They relate to basic organizational processes: designing, constructing, operating, maintaining, communicating, selecting, training, supervising and managing. This list is incomplete, but it will serve to show that they are the generic essentials of any productive process—the everyday business of management, in fact.

This brings us to two important generalizations. First, the quality of both production and protection is dependent upon the same underlying organizational processes. Safety is not a separate issue. Second, we cannot prevent latent conditions from being seeded into the system since they are an inevitable product of strategic decisions. All we can usefully do is to make them visible to those who manage and operate the organization so that the worst of them, at any one time, can be corrected. We cannot hope to solve all of our problems in one go. Resources are always limited. But we can target and then address the latent conditions most in need of urgent attention within a given period. Of course, while these are being fixed, other things will be going wrong. Risk management, like life, is 'one damn thing after another'. Some of the proactive diagnostic tools by which this long-term 'fitness programme' can be achieved will be described in Chapter 7.

In all three events, the essential process of checking and reviewing the defences broke down. In the train derailment, the underlying problem was virtually undiscoverable before the event. It is thus a very special case. More usually, latent weaknesses in defences are potentially—or even actually—evident prior to a bad outcome. The first two case studies revealed one very common reason why defensive weaknesses are not detected and repaired: the people involved had forgotten to be afraid.

It is very easy for those with production or profit goals uppermost in their minds to lose sight of the dangers. Bad events are mercifully rare in well defended systems. Very few people have direct experience of them. Production demands, on the other hand, are immediate, continuous and ever-present. They are also variable and attention-grabbing, whereas safe operations generate a constant—and hence relatively uninteresting—non-event outcome. The mechanisms by which this reliability is achieved can be opaque to those who operate and manage the system. Once again, Karl Weick has given us an elegant summary of this problem.[18]

> Reliability is invisible in the sense that reliable outcomes are constant, which means there is nothing to pay attention to. Operators see nothing and seeing nothing presume that nothing is happening. If nothing is happening and they continue to act the way they have been, nothing will continue to happen. This diagnosis is deceptive and misleading because dynamic inputs create stable outcomes.

Weick's point is that safety is a *dynamic* non-event—what produces the stable outcome is constant change rather than continuous repetition. To achieve this stability, a change in one system parameter must be compensated for by changes in other parameters.

If eternal vigilance is the price of liberty, then chronic unease is the price of safety. Studies of high-reliability organizations—systems having fewer than their 'fair share' of accidents—indicate that the people who operate and manage them tend to assume that each day will be a bad day and act accordingly. But this is not an easy state to sustain, particularly when the thing about which one is uneasy has either not happened, or has happened a long time ago, and perhaps to another organization. Nor is this Cassandra-like attitude likely to be well received within certain organizational cultures, as will be shown below.

Ron Westrum, a leading American industrial sociologist, has distinguished organizational cultures according to the way they deal with safety-related information.[19] He identified three types of culture—pathological, bureaucratic and generative—and their principal characteristics are summarized in Table 2.2.

Table 2.2   How different organizational cultures handle safety information

Pathological culture	Bureaucratic culture	Generative culture
• Don't want to know.	• May not find out.	• Actively seek it.
• Messengers (whistle-blowers) are 'shot'.	• Messengers are listened to if they arrive.	• Messengers are trained and rewarded
• Responsibility is shirked.	• Responsibility is compartmentalized.	• Responsibility is shared.
• Failure is punished or concealed.	• Failures lead to local repairs.	• Failures lead to far-reaching reforms.
• New ideas are actively discouraged.	• New ideas often present problems.	• New ideas are welcomed.

Westrum argues that organizations conducting potentially hazardous operations need *requisite imagination*—a diversity of thinking and imagining that matches the variety of possible failure scenarios. The possession of this requisite imagination not only characterizes high-reliability (or generative) organizations, its absence—as we saw in the first two case studies—features prominently in the developmental stages of an organizational accident.

With hindsight, it is nearly always possible to identify, prior to a disaster, the presence of warning signs which, if heeded and acted upon, could have thwarted the accident sequence. The question that often arises after the event is: How could these warnings have been missed or ignored at the time? There are a number of possible reasons why this happens, but most of them have to do with the fact that after-the-fact observers armed with '20/20' hindsight view events quite differently from the active participants who possessed only limited foresight. Knowing how events turned out—what psychologists have called *outcome knowledge*—profoundly biases our judgement of the actions of those on the spot. Several studies[20] have shown that:

- people greatly overestimate what they would have known in foresight,
- they also overestimate what others knew in foresight,
- they misremember what they themselves knew in foresight.

One of the facts we most readily overlook when we review the causal history of an accident is that some prior indication of disaster is only truly a warning if you know what kind of disaster you will suffer. But this is rarely the case, as the Dutch psychologists, Willem Albert Wagenaar and Jop Groeneweg, have explained:

> Accidents appear to be the result of highly complex coincidences which could rarely be foreseen by the people involved. The unpredictability is caused by the large number of causes and by the spread of information over the participants.... . Accidents do not occur because people gamble and lose, they occur because people do not believe that the accident that is about to occur is at all possible.[21]

In short, many accidents are *impossible accidents*—at least from the perspective of those involved.[22] This was probably the case for the aircraft maintenance accident, but perhaps it was not so true of the Barings Bank collapse. In that case, the Barings' management had been sufficiently alarmed by the fraud perpetrated by the Wall Street trader in April 1994 to review their own risk controls. They noted the deficiencies and started to correct them, but not quickly enough to avert the disaster. Perhaps, as their Group Treasurer observed, 'There was always something else that seemed more pressing'. That is an appropriate epitaph for most organizational accidents.

Notes

1 Sir Francis Bacon expressed this more elegantly in 1620: 'The human mind is prone to suppose the existence of more order and regularity in the world than it finds' (*The New Organon*). One means of simplification is to presume a symmetry of magnitude between causes and consequences.

2 See K.E. Weick, 'The vulnerable system: an analysis of the Tenerife air disaster' in P.J. Frost *et al.*, (eds), *Reframing Organizational Culture*, (London: Sage Publications, 1991). Karl Weick is one of the most perceptive of the social scientists writing about organizational accidents, and his work will be cited at various points throughout this book.

3 *Report on the Incident to Boeing 737-400, G-OBMM, Near Daventry on 23 February 1995*, Aircraft Incident Report 3/96, Department of Transport, (London: HMSO, 1996).

4 A borescope is a fibre-optic diagnostic device for detecting signs of structural defects in turbine blades within the high- and low-pressure areas of a jet engine. It forms an important part of the aircraft engineer's collection of non-destructive testing (NDT) techniques.

5 See *Report on the Accident to BAC One-Eleven, G-BJRT, over Didcot, Oxfordshire on 10 June 1990*, Aircraft Accident Report 1/92, Department of Transport, (London: HMSO, 1992). See also D. Maurino, J. Reason, N. Johnston and R. Lee, *Beyond Aviation Human Factors: Safety in High Technology Systems*, (Aldershot: Avebury Aviation, 1995), ch. 4.

6 *The Times*, 26 July 1996.

7   Ibid.

8   Nick Leeson's account of his part in the Barings collapse was entitled *Rogue Trader* (London: Little, Brown and Company, 1996).

9   Those seeking further information on the Barings collapse should read Stephen Fay's *The Collapse of Barings* (London: Arrow Business Books, 1996) from which the bulk of the story presented here was taken.

10 Baring Futures (Singapore) Pte Ltd, *The Report of the Inspectors Appointed by the Minister for Finance*, (Singapore: Ministry for Finance, 1995).

11 S. Fay, op. cit.

12 S. Fay, op. cit.

13 N. Leeson, op. cit.

14 S. Fay, op. cit.

15 S. Fay, op. cit.

16 *Report of the Board of Banking Supervision Inquiry into the Circumstances of the Collapse of Barings*, (London: HMSO, 1995).

17 I am most grateful to Peter Harle, Chief, Accident Prevention, Transport Safety Board (of Canada) for giving me a synopsis of this accident entitled *Nakina— An Organizational Accident*. A full account of the accident can be found in the TSB *Occurrence Report*, No. R92T0183, (Ottawa: Transport Safety Board, 1992).

18 K.E. Weick, 'Organizational culture as a source of high reliability', *California Management Review*, 29, pp. 112–27.

19 R. Westrum, 'Cultures with requisite imagination' in J. Wise, D. Hopkin and P. Stager (eds), *Verification and Validation of Complex Systems: Human Factors Issues*, (Berlin: Springer-Verlag, 1992), pp. 401–16.

20 Perhaps the most important of these is Baruch Fischhoff's paper 'Hindsight does not equal foresight: the effect of outcome knowledge on judgement under uncertainty', *Journal of Experimental Psychology: Human Performance and Perception*, 1, 1975, pp. 288–99.

21 W.A. Wagenaar and J. Groeneweg, 'Accidents at sea: multiple causes and impossible consequences', *International Journal of Man–Machine Studies*, 27, 1987, pp. 587–98.

22 See also Charles Perrow's highly influential book, *Normal Accidents: Living with High-Risk Technologies*, (New York: Basic Books, 1984).

3 Dangerous Defences

Killed by their Armour

On a damp late October morning in 1415, a considerable force of heavily armoured French cavalry advanced towards a small and sickly English army made up of 5 000 lightly clad archers and around 300 men-at-arms (the French army was at least five times this number). The mounted French knights rode across ground bounded on either side by thick woods. Although they had intended to attack the flanks of the English army to avoid the well understood threat of their longbows, the terrain caused the two wings of the attack to bunch together in the centre of the field. When they came within range, the English archers loosed a storm of yard-long steel-tipped arrows. Some of the French were killed outright; but many were thrown from their disabled horses.

By the early fifteenth century, the plate armour worn by knights had almost reached its zenith of weight and sophistication. It was proof against most penetrating and edged weapons, but it had a fatal flaw. The armour was so heavy that its unhorsed occupant found it difficult to get to his feet, particularly in the confined and muddy conditions that prevailed on the battlefield of Agincourt.[1] Once on the ground, they lay helpless and were slaughtered by the unencumbered English foot soldiers armed with mallets, spikes and daggers. (Some were taken prisoner and then killed.) While the English army lost around 100 men and boys in the battle, the French dead, mostly nobles, ran into many thousands.

This bloody episode in a long war—that the English eventually lost—serves to introduce the basic theme of this chapter, namely that defences designed to protect against one kind of hazard can render their users prey to other kinds of danger, usually not foreseen by those who created them, or even appreciated by those who use them. In short, defences can be dangerous. This is no less true now—in the age of high-technology systems—than it was at the time of Agincourt.

41

Some Paradoxes

The history of defences, barriers and safeguards abounds with para-doxes, often painful ones. Protective measures can cause harm. Conversely, small doses of a harmful entity can provide long-lasting protection, as in vaccination or inoculation.

A parallel to inoculation in the world of hazardous technologies is the opportunity to reform the system afforded by incidents, near-misses and other free lessons—events that could have been disastrous, but which were not on that occasion. To some degree, systems can be 'vaccinated' against organizational accidents by learning more about the strengths and weaknesses of their defences from these close shaves (see Chapter 6).

This chapter considers some of these defence-related ironies and paradoxes as they apply to systems that are vulnerable to organiz-ational accidents. At no point will it be argued that defences are intrinsically bad. There is no doubt that, in absolute terms, the pro-vision of redundant and diverse defences has greatly reduced the numbers of adverse events. They have, however, radically changed both the nature of the accidents that do happen and the character of the systems they protect—most particularly, they have transformed the relationship between the system's human and technical compo-nents. It is essential, therefore, that those who manage and operate complex technologies should appreciate both the advantages and the dangers of their multi-layered defences.

Automation: Ironies, Traps and Surprises

One of the ways in which system designers have sought to reduce what has come to be known as the 80:20 problem—the common finding that around 80 per cent of accident causes are due to human failures and only 20 per cent to technical failures—has been to pro-vide ever more automation at the human–system interface. Of course, the wish to distance fallible human beings from the control loop is not the only reason for this rapid increase in automation. A powerful incentive has been the availability of cheap computing power over the last 20 years, and the use of such leading edge technology can offer considerable commercial advantages.

When the European consortium, Airbus Industrie, was formed in the early 1970s, Europe's share of the world commercial jet transport market was close to zero. In order to compete with the dominant US manufacturers, Airbus resolved to use the latest available technology in the design of automated cockpits. Although other manufacturers such as Boeing and McDonnell Douglas followed suit, Airbus

Industrie continued to pioneer the division of labour between pilots and computerized flight management systems. Beating the competition required an aggressive policy of being different—particularly in the lengths to which they were prepared to go in order to make aircraft less reliant upon human control. Their radical philosophy of flight-deck automation gained them a substantial place in the jet transport market, but it also created a fresh crop of human factors difficulties. These problems are by no means unique to Airbus aircraft, or indeed to aviation.[2]

Before looking in detail at these aviation problems, it is worth considering in general terms what the British engineering psychologist, Lisanne Bainbridge, has aptly called 'the ironies of automation'.[3]

- By taking away the easy parts of a human operator's task, automation can make the difficult parts of the job even more difficult.
- Many systems designers regard human beings as unreliable and inefficient, yet they still leave people to cope with those tasks that the designer could not think how to automate—most especially, the job of restoring the system to a safe state after some unforeseen failure.
- In highly automated systems, the task of the human operator is to monitor the system to ensure that the 'automatics' are working as they should. But it is well known that even the best motivated people have trouble maintaining vigilance for long periods of time. They are thus ill-suited to watch out for these very rare abnormal conditions.
- Skills need to be practised continuously in order to preserve them. Yet an automatic system that fails only very occasionally denies the human operator the opportunity to practise the skills that will be called upon in an emergency. Thus, operators can become deskilled in just those abilities that justify their marginalized existence.
- And, as Bainbridge pointed out, 'Perhaps the final irony is that it is the most successful automated systems with rare need for manual intervention which may need the greatest investment in operator training'.

Although automatic control systems and their problems are not unique to aviation, it is the most extensively studied domain since it yields the best accounts of automation-related incidents and accidents—thanks to on-board flight data recording. In addition, flight management systems represent some of the most advanced forms of automation currently available. For these reasons—and because many of the problems found in aircraft are common to other technologies—we will focus mainly upon aircraft automation in the remainder of this section.

Flight deck automation has proved to be a mixed blessing. On the one hand, it has greatly simplified the task of horizontal navigation—knowing where you are in relation to the ground—through an electronic map that not only shows the aircraft's position relative to the geography beneath, but displays en route weather as well. Pilots regard this as one of the most useful features of 'glass-cockpit' aircraft. On the other hand, there have been many problems with height changes or vertical navigation, some of them leading to fatal accidents.[4]

Vertical navigation is complex because it is controlled by a combination of elevator and engine thrust inputs. A Flight Management System (FMS) offers at least five ways for changing altitude, each with different levels of automation. Some of these modes will be selected by the pilot, others will come into play automatically. In one mode, for example, the FMS will control the aircraft's vertical speed; in another, it will maintain a given flight path angle. A further complication is that the FMS will automatically switch modes according to the local situation.[5] For example, if the pilot had selected a target height of 5000 feet, the climb mode would remain active only until this altitude is reached. At that point, the aircraft would level off and the FMS would automatically switch to the 'altitude hold' mode.

In addition, some flight management systems are fitted with 'hard' protection envelopes. These are designed to prevent the aircraft's speed from going outside certain critical values, and from exceeding the airframe's stress tolerances. The term 'hard' means that the FMS will impose these barriers automatically without the need for pilot input. The idea is that should the pilot erroneously take the aircraft beyond these 'hard' protection limits, the FMS will switch itself into a 'safe' mode setting. If, for example, an aircraft's speed dropped below a certain critical value on the approach, the FMS could initiate the 'go around' mode by automatically advancing the throttles. This defensive feature has proved dangerous on a number of occasions when pilots, unaware of a mode transition, have fought with the FMS for control of the aircraft—sometimes unsuccessfully.

The nub of the problem—from a human factors standpoint—is that the FMS can change modes either as the result of a pilot intervention or according to its own internal logic. It is not surprising, therefore, that pilots can become confused about which mode is active, or what will happen next. In order to keep 'ahead' of the aircraft, flight crews must not only know which mode they are in, but also how the FMS selects new modes. This is sometimes a tall order, particularly when these demands fall, as they often do, during periods of high mental workload. Here then is another irony of automation: flight management systems designed to ease the pilot's mental burden tend to be most enigmatic and attention-demanding during periods of maximum workload—a feature that Earl Wiener, an aviation psychologist at the University of Miami, has termed 'clumsy automation'.[6]

A recent study at the Massachusetts Institute of Technology examined 184 cases of mode confusion—some resulting in fatal accidents—occurring between 1990 and 1994.[7] Of these, 74 per cent were associated with vertical navigation manoeuvres. Only 26 per cent were related to horizontal navigation. Mode errors were classified into a number of categories. The largest class (45 per cent of the total) was due to pilots entering data incorrectly or into the wrong mode, which resulted in an unexpected FMS mode or a surprising mode change. The second largest category (20 per cent) was associated with mode transition problems in which pilot confusion (or an unplanned aircraft deviation) occurred because the FMS executed an unexpected mode transition or failed to perform an expected one. A slightly smaller group of mode errors (18 per cent) arose because the flight crew did not fully understand the automation and consequently either made an inappropriate input—or failed to make a necessary one—creating confusion on their part or an unexpected aircraft manoeuvre. Of these confusions, 14 per cent were due to a breakdown in crew coordination—a sequence of events initiated by one crew member caused others on the flight deck to make wrong assumptions about a mode, a mode transition or a deviation. And, in 12 per cent of cases, a failure on the part of the FMS created confusion or an unwanted aircraft deviation.

These findings reveal yet another irony of automation. Computerized control systems intended to remove the possibility of localized slips, lapses and fumbles on the flight deck can increase the probability of higher-level mistakes with the capacity to cause the destruction of the entire aircraft and its occupants. While the mental burdens of horizontal navigation have been greatly reduced, those relating to the more complex and safety-critical task of vertical navigation can, under certain circumstances, be markedly increased. Furthermore, these additional problems involving height changes will occur during takeoffs and landings—the periods of greatest pilot workload.

In their detailed study of mode confusions (also known as 'mode errors') in a variety of automated control systems, including aviation, anaesthesia, nuclear power plants and space flight,[8] David Woods and Nadine Sarter have identified a number of human factors problems that apply across all of these domains. Automated systems can:

- increase demands on users' memory,
- cause users to be uncertain as to where and when they should focus their attention,

- make it difficult for users working in teams to share the same situational awareness,
- impair mental models of the system,
- increase workload during high-demand periods,
- limit the users' ability to develop effective strategies for coping with task demands,
- increase stress and anxiety,
- increase the potential for confusion through enhanced flexibility (i.e., many possible levels and types of automation).

Woods and Sarter go on to identify the properties of automated systems that promote these problems:

- they can hide interesting events, changes and anomalies,
- they possess multiple modes,
- they force serial access to highly related information,
- they offer a 'keyhole' view of a limitless virtual space,
- they contain complex and arbitrary sequences of operations and modes,
- they suppress important information about the activities of other team members.

In summary, mode confusions can occur for one of two main reasons: either the user makes a wrong assessment of the active mode at a particular time or the user fails to notice transitions in mode status. The first is a failure of perception (or interpretation), the second is a failure of attention. In both cases, the causes can be traced to 'clumsy' automation. In their efforts to compensate for the unreliability of human performance, the designers of automated control systems have unwittingly created opportunities for new error types that can be even more serious than those they were seeking to avoid.

Quality Control Versus Quality Assurance

Has the shift from quality control to quality assurance opened a gap in the protection of hazardous technologies? To understand this question, we need to review very briefly the milestones in the history of achieving and measuring quality.[9]

The ancients—Babylonians, Egyptians, Greeks and Romans—initiated the quality process by establishing the units of weight and dimension. Throughout the Middle Ages, trade and craft guilds set standards for quality as well as for working conditions and wages. With the arrival of the Industrial Revolution in Britain, however, and with the subsequent spread of mass production methods in the United

States, making things was no longer the exclusive province of experienced craftsmen. Jigs, templates and moulds combined with powered machinery meant that complex manufacturing work could be carried out by relatively untrained people. During this period the main responsibility for monitoring quality fell to the supervisors. Later, following the impact of Taylorism (scientific management) and a substantial increase in worker-to-supervisor ratios, supervisors concentrated more on production issues and left the control of quality to a new breed of overseer, the inspector. The term used to describe this specialist inspection function at the end of the production line was *quality control*.

Working at Bell Laboratories in New York in the 1920s, Walter Shewhart and his team invented Statistical Process Control (SPC). A crucial feature of this technique was that it required quality measurements to be made at the point of manufacture rather than at the end of the line. SPC was later adopted by W. Edwards Deming in the 1950s and taken to Japan. The following passage describes the situation, as Deming found it, in post-war Japan:[10]

> In management, Japan also lagged behind, using the so-called Taylor method in certain quarters... . Quality control was totally dependent on inspection, and not every product was sufficiently inspected. In those days, Japan was still competing with cost and price, but not with quality. It was literally still the age of 'cheap and poor' products.

Deming developed the Plan–Do–Check Action Cycle to guide continuous quality improvement. The general tendency in an action-oriented society is to focus on the 'doing' part of the cycle and to skimp on the planning and checking phases.

In the 1970s, Deming re-imported these ideas to the United States. Other quality 'gurus', like Joseph Juran, Armand Feigenbaum and Kaoru Ishikawa, were also having a profound effect upon the beliefs and attitudes of American managers. The outcome was the Total Quality Management (TQM) movement that has subsequently spread throughout the industrialized world. A key feature of TQM is that everyone in a company shares the responsibility for quality. Quality was not something to be 'controlled' at the end of the line by inspectors, but something that had to be 'assured' throughout the entire work process, from top management decisions to the individual actions on the shopfloor. Hence, we now speak of *quality assurance* rather than *quality control*. Quality, in other words, can only be 'engineered' into a task, not 'inspected' into it.

So much for the background. Now let us see what effect this development has had upon one particular activity—aircraft maintenance —where quality and safety are inextricably linked. It must be stressed, however, that the discussion below is relevant to all work domains in which modern quality assurance techniques are being used, and where the quality of work has a direct impact upon system safety.

In Chapter 2, we examined the case of the missing engine rotor drive covers that led to a near-disaster. In his report, the accident investigator noted a number of similarities between the precursors of this event and those of other British maintenance-related accidents. Two common features in these incidents were, first, that a senior engineer 'signed off' his own work as being satisfactory when it was not; and, second, no member of the Quality Assurance department was available during night-time hours to assess the adequacy of working practices.

In the absence of a separate inspection, a suitably authorized individual must sign or place his personal stamp against a declaration that the work has been performed to a high standard of quality. This represents the only 'proof' that the job has been completed, and a full set of such stamps or signatures is required to show that an aircraft is fit to fly following maintenance work.

Are such quality assurance (QA) measures a sufficient guarantee of the airworthiness of an aircraft? These incidents suggest that they are not. The lack of a separate inspection plus the absence of QA staff allowed unfinished or wrongly executed jobs to be signed off as fit for service.

One important lesson to be drawn from these events is that we need to be cautious about defensive developments designed to improve—or at least ensure—quality and safety when they can also serve other goals, unrelated or even inimical to quality and safety. Let us consider, for a moment, what else QA offers to both the organization and the individual. From a company perspective, the abolition of an independent inspection function (for most jobs, at least) confers at least two commercial advantages. First, it removes the often lengthy delays associated with separate inspections. Second, it allows more of the workforce to be engaged in obviously productive 'doing' work as opposed to 'checking' work.

At an individual level, QA practices offer the hard-pressed engineer a path of least effort: namely, to sign off on task steps—either before or after the event—without actually monitoring the quality of the work. One of the enduring findings of work psychology is that people will be tempted to take short-cuts whenever such opportunities present themselves. It is no accident, therefore, that 'signing off without checking' is one of the more common procedural violations to be found in aircraft maintenance.

It is not being suggested that either the organization or the individual is *deliberately* cheating on quality or safety. Indeed, they are both likely to declare their genuine commitment to such goals, and point to their adoption of the latest quality assurance techniques as clear evidence of this. In a continually evolving discipline like aircraft engineering it is hard to appreciate that not all changes are for the better. While the emperor's new clothes may not have left him entirely naked, he could well have been stripped of an important garment.

Working life is very complicated. We are subject to many forces, not all of which are pushing in the same direction at the same time. As we have seen before, conflicts between production and protection pressures tend to be resolved in favour of the former—at least until a bad accident occurs.

Writing Another Procedure

All organizations suffer a tension between the natural variability of human behaviour and the system's needs for a high degree of regularity in the activities of its members. The managers of hazardous systems must try to restrict human actions to pathways that are not only efficient and productive, but also safe. The most widely used means to achieve both goals are written procedures. But there are a number of important differences between the procedures for production and those for protection.

Although by no means immutable, the procedures designed to ensure efficient working tend to arise fairly naturally from the nature of the productive equipment and the task to which it is put. Safe operating procedures, on the other hand, are continually being amended to prohibit actions that have been implicated in some recent accident or incident.[11] Over time, these additions to the 'rule book' becoming increasingly restrictive, often reducing the range of permitted actions to far less than those necessary to get the job done under anything but optimal conditions.

Figure 3.1 illustrates this shrinkage of allowable action as it occurs over the history of a given system. This could be a chemical process plant, a railway, an aircraft operating company—or any hazardous technology at risk to organizational accidents. The space between the shaded areas represents the scope of prescribed action. As time passes, the organization inevitably suffers accidents and incidents in which human actions are identified as contributing factors. After each event, the procedures are modified so as to proscribe these implicated actions. As a consequence, the scope of allowable actions gradually shrinks to a range that is less than that required to perform all the necessary tasks. The only way to do these jobs is to violate the procedures.

For example, the British Rail Rule Book prohibits shunters (the people who join up the wagons and carriages that go to form a train)

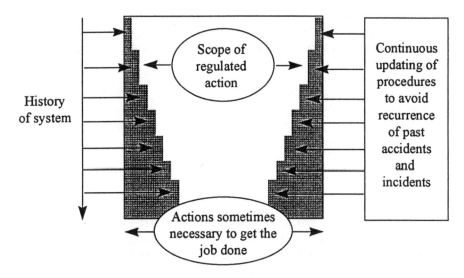

Figure 3.1   **How necessary additional safety procedures reduce the scope of action required to perform tasks effectively**

from remaining between wagons during easing up—that is, when a set of wagons are being propelled by a pilot engine towards some stationary wagons to which they need to be attached.[12] Only when the wagons are stopped can the shunter go between them to make the necessary coupling. Sometimes, however, the shackle for connecting the wagons is too short to be coupled when the buffers are at their full extension. The job can only be done when they are momentarily compressed as the wagons first come into contact. The best way of making this connection is by remaining between the wagons during the easing up process. Shunting is a dangerous job. Many shunters die as the result of being caught between the buffers or by falling under the wheels. Just the act of violating the Rule Book does not kill them. There is adequate clearance between both the buffers and the wagons. What sometimes proves fatal is the subsequent error that is committed while the shunter is violating the Rule Book by being between the wagons. He could be momentarily distracted and unwittingly place himself between a set of buffers, or could slip and fall under the wagon wheel. A violation plus an error is frequently a formula for disaster in hazardous work.[13]

However, the mere repetition of these accident-implicated actions in isolation does not usually bring about a bad outcome. Their contribution to an earlier event was merely one cause—necessary, perhaps, but hardly ever sufficient—among a complex interaction of latent conditions, local triggers and other active failures. Because this combination of factors rarely, if ever, recurs in precisely the same form, the workforce quickly learns that these isolated violations generally carry no penalty. Indeed, they often discover that they lead to an easier and more efficient way of working.

Ironically, then, one of the effects of continually tightening up safe working practices is to increase the likelihood of violations being committed. The scope of permitted action shrinks to such an extent that procedures are either violated routinely, or on those occasions when operational necessity demands it. Whereas errors arise from the underspecification of various mental operations, many violations are created by procedural overspecification.[14]

But that is not the end of the story. Although the commission of these isolated forbidden acts may be mostly inconsequential, the fact that they *were* implicated in some past event indicates that they *can* increase the risks for their perpetrators, even though they do not usually have a bad outcome.

One of the ways in which violations increase risk is by making a subsequent error more likely to happen. Driving a car at 110 mph, for example, is not in itself sufficient to cause an accident. However, since this speed will almost certainly be higher than the driver's norm, there is a greater chance of misjudging the braking distance or the vehicle's handling characteristics. Since the margins for tolerating such errors are now considerably reduced, the error is less likely to be forgiven. Violations can thus have two consequences. They can increase the probability of a later error, and they can also increase the likelihood that it will have a bad outcome.

The important issue in many hazardous technologies is not *whether* to violate, but *when* to violate—or perhaps, more importantly, when to comply. When proscribed actions are necessary in order to get the job done, then the rules will be violated. Nearly all hazardous operations involve making actions that lie outside the prescribed boundaries yet remain within the limits of what would be judged as acceptable practice by people sharing comparable skills. Most experienced workers know approximately where the 'edge' between safety and disaster lies and do not exceed it, except in extreme circumstances. What they do not always appreciate, however, is where they currently are in relation to that edge. There are times when it is prudent to retreat within the narrow boundaries of permitted action or even to stop operations altogether. Deciding when to do this, though, is a matter of delicate and sometimes fallible judgement.

Causing the Next Accident by Trying to Prevent the Last One

Accident investigators are required not only to establish the causes of an event but also to recommend measures that will help to prevent its recurrence. Often, these preventative measures take the form of engineering or regulatory 'fixes' designed to overcome a particular problem or systemic weakness that featured conspicuously among the causal factors. Unfortunately, these very same 'fixes' sometimes play a major part in causing some subsequent accident. Two examples will illustrate this point.

Following a number of air accidents in the 1950s in which aircraft crashed on takeoff from slushy or otherwise contaminated runways, a new flight deck instrument was introduced—the takeoff monitor.[15] This directed the pilot to adjust the aircraft's angle of climb to match the position of a target marker shown on the instrument. The pilot manipulated the throttle and elevator controls until a cursor—indicating the aircraft's current climb angle—coincided on a vertical scale with the target marker. It was a relatively novel 'command instrument' that calculated the appropriate climb angle for the current conditions and then 'ordered' the flight crew to match this target angle by tracking the aircraft cursor up the scale.

Shortly after the introduction of the takeoff monitor, an aircraft adopted an unusually steep angle of climb on takeoff.[16] Unable to sustain it, the aircraft stalled and crashed back on to the runway. The accident investigators subsequently found that a small retaining screw had failed within the takeoff monitor, causing it to 'command' a wholly inappropriate angle of climb. Despite the fact that their remaining flight instruments were intact, the pilots obeyed the faulty indicator. As discovered elsewhere, instruments that do some of the pilots' thinking for them have a way of capturing attention—the problem was termed 'fascination' at the time[17]—even though alternative, but less easily processed, sources of information are available on the flight deck.

The second example begins with a 'sentinel event' in the history of nuclear power generation. At 0400 on 28 March 1979 a turbine in one of Metropolitan Edison's two pressurized water reactors (PWRs) on Three Mile Island—near Harrisburg, Pennsylvania—stopped (tripped) automatically.[18] The cause was the leak of a cupful of water through a faulty seal during maintenance work. The moisture interrupted the air pressure applied to two valves connecting to the feedwater pumps and caused them to shut down automatically. This, in turn, cut the water flow to the steam generator and tripped the turbine. Without the pumps working, the heat of the primary cooling system—circulating through the reactor core—could not be transferred to the cooler water in the secondary circuit. At this point, hundreds of alarms and annunciators came on in the control room and the emergency feedwater pumps started up, again automatically. These were designed to pull water from an emergency storage tank and run it through the secondary cooling system to compensate for the water that boils off once it is not circulating.

The emergency lasted for 16 hours and resulted in the release of small quantities of radioactive material into the surrounding atmosphere. The operators on duty at the start of the event made a number of well documented errors, but the one that most concerns us here was the cutting back of the high-pressure injection of water into the reactor coolant system. This reduced the net flow rate from around 1000 gallons per minute to about 25 gallons per minute. This 'throttling' caused serious damage to the reactor's core. It is not necessary for us to go into the reasons why this action was taken—they have been discussed at length elsewhere. What concerns us are the measures taken by the United States Nuclear Regulatory Commission to prevent the recurrence of this step in any future nuclear power plant event.

Following Three Mile Island (TMI), the US nuclear regulators made it a requirement that operators should not cut back on the high-pressure injection during the recovery stages of an off-normal event. Three years later, a control room crew at another PWR, this time at Ginna,[19] were faced with an emergency in which reducing the high-pressure injection would have been an appropriate step to take. Well aware of the post-TMI regulatory restriction on this action, they did not take it. As a result, the emergency was prolonged by several hours.

These examples illustrate some of the potential dangers that can ensue from what, on the face it, appear to be perfectly sensible attempts to 'fix' the causes of previous accidents. There are two main problems. As the case of the takeoff monitor showed, introducing new engineered defensive features adds complexity to the system. In particular, it adds components which themselves can fail. Moreover, in this case as in the more modern varieties of automation, attempts to distance pilots from the direct control loop can create unforeseen types of human error.

The second problem, illustrated by the Ginna event, is that regulators—just as much as system designers—cannot foresee all the possible scenarios of failure in complex, tightly-coupled and highly interactive systems such as nuclear power plants, and so cannot universally proscribe particular types of human response. What proved to be an error in the TMI event turned out to be a vital step at Ginna. As we shall see in Chapter 4, regulations and procedures share with other feedforward control devices the problem of being insensitive to local conditions.

Defences-in-Depth: Protection or Dangerous Concealment?

The philosophy of defences-in-depth is summarized in Figure 3.2. Here, safety depends upon the *causal independence* between errors and technical faults affecting the process equipment, the normally inactive engineered protective systems, the physical barriers against unplanned release of dangerous substances and the proximity of the likely victims to the bad event. Major accidents occur when this assumption of mutual independence between the various layers of defence is violated. During the Bhopal disaster, for example, three supposedly independent defences failed simultaneously: a flare tower to burn off the deadly methocyanate gas, a scrubber to clean air emissions and a water sprinkler system to neutralize the remaining fumes.[20]

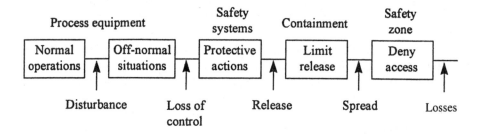

Figure 3.2   Summarizing the philosophy underlying defences-in-depth

The horizontal lines between the boxes indicate the consequences of failure at each stage. For the defensive layers to maintain safety, it is necessary that failures affecting one box will be unrelated to failures affecting other boxes (after Rasmussen).

Defences-in-depth are thus built upon redundancy (many layers of protection) and diversity (many different varieties of protection). However, it is these very features—which are highly desirable from an engineering standpoint—that also create a variety of problems in complex sociotechnical systems—to such an extent that Jens Rasmussen, engineer and leading philosopher of technology, has coined the phrase 'the fallacy of defences-in-depth'.[21]

One of the main problems that defences-in-depth pose for a system's managers and controllers is that they can conceal both the occurrence of their errors and their longer-term consequences. A characteristic of such defences is that they do not always respond to individual failures. These can either be countered or concealed, and in neither case need the individuals directly concerned be aware of their existence. This feature allows for the insidious build-up of the latent conditions—resident pathogens—that may subsequently combine with local conditions and 'sharp-end' errors to breach or bypass the defensive layers.

A system with few defences—driving a car, for example—does not normally remain so impassive in the face of accumulated failures. Mistakes are immediately detectable and the subsequent learning will limit their recurrence. This does not necessarily happen with multi-layered defences. A dangerous penalty of this concealment is that neither individuals nor organizations can easily profit from their errors. This penalty is even more likely to be exacted when those who design or manage such a system are more inclined to attribute its occasional catastrophic failures to individual fallibility than to intrinsic system weaknesses.

In 1984 Charles Perrow, an organizational theorist, published a highly influential book, *Normal Accidents: Living with High-Risk Technologies,* in which he advanced the bleak proposition that accidents are inevitable in complex, tightly-coupled systems like nuclear power plants—regardless of the skills of their operators and managers.[22] Hence the title: accidents in such systems are 'normal'. According to Perrow, the redundancies that go to make up defences-in-depth have three dangerous features.

- Redundant defensive back-ups increase the interactive complexity of high-technology organizations and thus increase the likelihood of unforeseeable common-mode failures. While the assumption of independence may be appropriate for purely technical breakdowns, human errors at the 'sharp end', in the maintenance sector and in the managerial domains are uniquely capable of creating failures that can affect a number of defensive layers simultaneously.
- Adding redundancy makes the system more opaque to the people who nominally control and manage it. Undiscovered errors and other latent problems accumulate over time and thus increase the likelihood of the 'holes' in the defensive layers lining up to permit the passage of an accident trajectory (see Figure 1.5). This alignment of the gaps can be created either by interactive common-mode failures or by the simultaneous disabling of supposedly independent defences, as at Chernobyl.
- As a consequence of this dangerous concealment, and because of their obvious engineering sophistication, redundant defences

can cause system operators and managers to forget to be afraid. This false sense of security prompts them to strive for even higher levels of production (see Chapter 1). As Perrow put it: 'Fixes, including safety devices, often merely allow those in charge to run the system faster, or in worse weather, or with bigger explosives.'

False Alarms

To those who have been enraged by the repeated wailing of a neighbour's car alarm while the vehicle stands undisturbed, the idea that alarms can lie will come as no surprise. In everyday life these false alarms can create intense irritation. In hazardous technologies they can cause disaster. It is the 'cry wolf' situation. Frequent false alarms cause people to lose trust, and to ignore or disbelieve warnings when they signal genuine emergencies.

At just after 4.00pm on Sunday, 18 June 1972, a British European Airways Trident passenger aircraft (Papa India) took off from London Heathrow for Brussels and crashed 120 seconds later in a field on the outskirts of Staines.[23] There were no survivors. This was the worst air disaster to occur within the British Isles and the first fatal accident involving a Trident in normal airline operations. It also remains one of the most mysterious. Although an intact Flight Data Recorder (FDR) revealed exactly *what* happened to the aircraft, we have no certain knowledge of *how* it happened or *why* it happened, since there was no Cockpit Voice Recorder (CVR) aboard at that time.

Unusually, there were four pilots on the flight deck. A senior, 51-year-old BEA captain was in the left-hand seat, a 22-year-old junior pilot was in the right-hand seat, and behind and between them was a somewhat more experienced 24-year-old pilot whose task was to monitor the instruments and the performance of the other two. Further back, in the jump seat, was another BEA captain. The action that initiated the disaster is not in dispute. At around one-and-a-half minutes into the flight, the Trident's droops, mounted on the leading edges of the wings, were retracted prematurely at too low an airspeed and the aircraft entered an incipient stall. We do not know who moved the control or why it was done.

The Trident was fitted with a number of stall warning systems. In addition to audio-alarms, there was also a 'stick-shaker' warning system and a pneumatically-powered 'stick pusher' stall recovery system. The latter automatically disengages the autopilot and lowers the nose of the aircraft in order to recover from the incipient stall. There was also a lever on the left side of the central control pedestal that could disable the stick-shaker warning system.

The FDR evidence revealed that both the stick-shaker and the stick-pusher mechanisms came into play as soon as the aircraft approached the stalling speed. The aircraft's nose pitched down, but with the elevator trim unadjusted, the aircraft had become tail heavy. As a result the nose pitched up again and eight seconds after the first stick push, the stall recovery system operated a second time. Three seconds later, the same pattern repeated itself, and once again the stick-pusher forced the nose down. At this stage, someone on the flight deck made a fatal error. The stick warning mechanism was turned off. The nose of the aircraft pitched up in excess of 30 degrees and entered a true stall. It then descended vertically and hit the ground 22 seconds later.

Why was the stick-shaker mechanism disabled on its third activation? We will never know for sure, but it is widely believed among the British commercial pilot community that the crucial reasons were, first, that the flight crew did not believe they were in a stalled state (probably because they were unaware of the droop retraction), and, second, they profoundly mistrusted the stick-shaker warning system. In short, they thought it was indicating falsely. There had not only been several false alarms in the development stages of the recovery system some years earlier but also a number of documented incidents involving the false activation of this system during line operations.

Deliberate Weak Links

One way of limiting the damage caused by an accident or a technical breakdown is to design deliberate points of weakness into a system so that it will fail in safer and more predictable ways. The fuses in an electrical circuit are a commonplace example of this principle. Should a circuit become overloaded, it will 'blow' at the fuse wire rather than within an appliance, thus minimizing the risk of a fire breaking out in some unattended location.

A similar principle is used in the construction of large commercial aircraft whose engines are suspended beneath the wings by pylons. In a typical Boeing 747, for example, each engine pylon contains two kinds of fuse pin (or strut): an upper link fuse pin and an aft diagonal brace fuse pin. If an undue force is applied to the engine and its pylon, the assembly is designed to break away cleanly at the fuse pins, thereby avoiding the possibility of tearing off large pieces of the wing structure. An aircraft will survive a lost engine a good deal better than losing a substantial portion of its lift-producing structure, the wing.

While fuse pins may have served their intended function in the early days of suspended jet engines, their unwanted failure has recently been implicated in a number of serious accidents and incidents, most notably at Schiphol in 1992.[24] Quite often, these fuse pin failures have been due to faulty installation or improper fastening during maintenance work.

On 1 March 1994, a Northwest Airlines (NWA) Boeing 747-251B dropped and then dragged its Number 1 engine during the landing rollout at New Tokyo International Airport, Narita.[25] The immediate cause of this accident was the fracture of the upper link fuse pin within the engine pylon. This, in turn, was due to the failure(s) to secure the fastenings of the aft diagonal brace fuse pin, causing it to come free some time after a major overhaul check at NWA's Minneapolis/St Paul maintenance facility. The engine had been opened at that time to permit the non-destructive testing of the various fuse pins. The appropriate fastenings were later found in the hangar in an unmarked cloth bag concealed by a piece of wooden board on an underwing workstand.

This accident illustrates how the measures taken to avoid one safety problem can contribute to another. The migration of fuse pins from B-747 engine pylons had been reported on five occasions prior to the NWA Narita accident. One had resulted in an accident similar to that at Narita. All were attributed to improper installation—though design defects must surely have played their part. These incidents led Boeing to require the addition of secondary fuse pin retainers. At the time of the accident, seven of NWA's fleet of 41 B-747s had these secondary retainers installed on the aft diagonal braces. It is likely that this variety of fastenings contributed to the misunderstandings and procedural confusions that subsequently allowed the installation errors to go undetected.

Safety devices, like the addition of any other component, increase the complexity of a system, particularly during maintenance. As such, they create additional opportunities for human failure of an especially safety-critical kind. Once again, then, we have an instance of a safety feature becoming a point of vulnerability. The main hazard facing an engine in a modern aircraft is not so much excessively high loading as an excessive amount of contact with the human hand.

Summary

In this chapter we have identified at least six ways in which defences might be dangerous:

- Defensive measures designed to reduce the opportunities for a particular kind of human error can relocate the error opportunities to some other part of the system, and these errors may be even more costly.

- Gains in defences are often converted into productive, rather than protective, advantage, thus rendering the system less safe than it was before.
- Defences-in-depth, based upon redundancy and diversity, make the system more opaque to its operators, and hence allow the insidious build-up of latent conditions.
- Warnings and alarms that acquire a reputation for indicating dangers where none exist are less likely to be acted upon in the event of a true emergency.
- Measures designed to eliminate a conspicuous cause of some previous accident can contribute to the next one.
- Defences, barriers and safeguards add additional components and linkages. These not only make the system more complex, they can also fail catastrophically in their own right.

Notes

1 To find out more about what it felt like to be on the receiving end of an arrow storm, try John Keegan's *The Face of War* (New York: Viking Press, 1976).
2 Much of the information in this section came from the issue of *Aviation Week and Space Technology*, Part 1, devoted to 'Automated cockpits: who's in charge?', 30 January 1995.
3 L. Bainbridge, 'Ironies of automation' in J. Rasmussen, K. Duncan and J. Leplat (eds), *New Technology and Human Error*, (Chichester: Wiley, 1987), pp. 271–83.
4 D. Hughes, 'Incidents reveal mode confusion. Automated Cockpits Special Report, Part 1', *Aviation Week and Space Technology*, 30 January 1995, p. 5.
5 D.D. Woods, 'The price of flexibility' in W. Hefley and D. Murray (eds), *Proceedings of International Workshop on Intelligent User Interfaces*, ACM, January 1993. See also N.B Sarter and D.D. Woods, 'Mode error in the supervisory control of automated systems' in *Proceedings of the Human Factors Society 36th Annual Meeting*, (Atlanta, GA, October 1992).
6 E.L. Wiener, *Human Factors of Advanced Technology ('Glass Cockpit') Transport Aircraft*, Technical Report 117528, (Washington, DC.: NASA, 1989).
7 The work was carried out by R. John Hansman, an MIT professor in the Aeronautics and Astronautics Department. The data given here are from D. Hughes, op. cit.
8 N.B. Sarter and D.D. Woods, op. cit. See also D.D. Woods *et al.*, *Behind Human Error: Cognitive Systems, Computers and Hindsight. State of the Art Report*, (Dayton, Ohio: CSERIA, Wright-Patterson Airforce Base, 1994).
9 A. Bendell, J. Kelly, E. Merry and F. Sims, *Quality: Measuring and Monitoring*, (London: Century Business, 1993).
10 Quoted by Bendell *et al.* op. cit. p. 47. Originally from Nancy R. Mann's *The Keys to Excellence: The Story of the Deming Philosophy*, (London: Mercury, 1985).
11 J. Reason, D. Parker and R. Free, *Bending the Rules: The Varieties, Origins and Management of Safety Violations*, (Leiden: Faculty of Social Sciences, University of Leiden, 1994).
12 R. Free, *The Role of Procedural Violations in Railway Accidents*, PhD Thesis. University of Manchester, 1994.
13 Ibid.

14   J. Reason, 'A systems approach to organizational errors', *Ergonomics*, **38**, 1995, pp. 1708–21.
15   Conspicuous among these accidents was the crash of a BEA aircraft in February 1958 with, among others, the Manchester United football team and its manager, Sir Matt Busby, on board.
16   J. Reason, *Man in Motion*, (London: Weidenfeld, 1974).
17   FPRC, *Fascination*, Flying Personnel Research Committee Report, (London: Ministry of Defence, 1960).
18   J. Kemeny, *The Need for Change: The Legacy of TMI. Report of the Presidential Commission on the Accident at Three Mile Island*, (New York: Pergamon, 1979). See also C. Perrow, *Normal Accidents: Living with High-Risk Technologies*, (New York: Basic Books, 1984).
19   D.D. Woods, *Operator Decision Behavior during the Steam Generator Tube Rupture at the Ginna Nuclear Power Station*, Research Report 82-1C57-CONRM-R2, (Pittsburgh, Penn: Westinghouse R & D Center, 1982).
20   D.A. Lihou and S.J. Lihou, *Bhopal: Some Human Factors Considerations*, (Birmingham: Lihou Loss Prevention Services, 1985). See also L.J. Bellamy, *The Safety Management Factor: An Analysis of the Human Error Aspects of the Bhopal Disaster*, (London: Technica Ltd., 1985).
21   J. Rasmussen, 'Learning from experience? Some research issues in industrial risk management' in B. Wilpert and T. Qvale (eds), *Reliability and Safety in Hazardous Work Systems*, (Hove: LEA, 1993).
22   Ibid.
23   M. Job, *Air Disaster*, Vol. 1, (Weston Creek, ACT: Aerospace Publications Pty Ltd, 1994), ch.11, pp. 88–97.
24   'B-747 Accident, Schiphol Airport, Amsterdam, 4 October 1992', *Flight Deck*, **21**, (London: British Airways Safety Services, Autumn 1996), pp. 13–22.
25   NTSB, *Maintenance Anomaly Resulting in Dragged Engine during Landing Rollout, Northwest Airlines Flight 18, New Tokyo International Airport, March 1, 1994*, NTSB/SIR-94/02, (Washington, DC: National Transport Safety Board, 1995).

4   The Human Contribution

The Human Factor

People design, build, operate, maintain, manage and defend hazardous technologies. It is hardly surprising, therefore, that the human factor plays the major part in both causing and preventing organizational accidents. This chapter explores the human contribution to the safety—or otherwise—of complex well defended systems.

It has become fashionable to claim that human error is implicated in 80–90 per cent of all major accidents. While probably close to the truth, this statement adds very little to our understanding of how and why organizational accidents happen. In the first place, it could hardly be otherwise, given the range of human involvement in hazardous systems. Second, the term 'human error' conveys the impression that all unsafe acts can be lumped into a single category. But errors take different forms, have different psychological origins, occur in different parts of the system and require different methods of management.

In addition, this assertion fails to acknowledge that people's behaviour within hazardous systems is far more constrained than it is in everyday life. The actions of pilots, ships' crews, control room operators and the like are tightly governed by managerial and regulatory controls. These administrative controls form a major part of any hazardous system's defences and are of two main kinds.[1]

- *External controls* made up of rules, regulations, and procedures that closely prescribe what actions may be performed and how they should be carried out. Such paper-based controls embody the system's collective wisdom on how the work should be done.
- *Internal controls* derived from the knowledge and principles acquired through training and experience.

External controls are written down and need to be close at hand when people are carrying out the prescribed activities. Internal controls, on the other hand, exist mainly within an individual's head. Any attempt to classify organizational behaviours must begin by considering how various combinations of administrative controls work to confine the natural variability of human action to safe and productive pathways.

The Varieties of Administrative Controls

Administrative controls range from the mainly prescriptive to the largely discretionary. The former depend primarily upon external procedures, while the latter are provided by internalized knowledge and experience—or, in a word, training. Between the two are various blends and mixtures. This continuum of control modes is summarized in Figure 4.1.

A mainly prescriptive control mode is shown in Figure 4.2. Figure 4.3 illustrates the opposite end of the dimension, being predominantly discretionary. Figure 4.4 shows a mixture of control modes.[2]

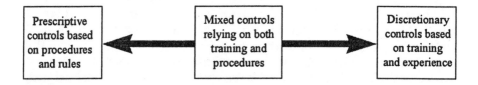

Figure 4.1   A continuum of administrative controls

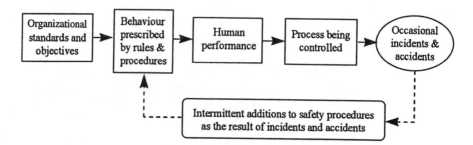

Figure 4.2   A mainly feedforward process control based on procedures with intermittent additions

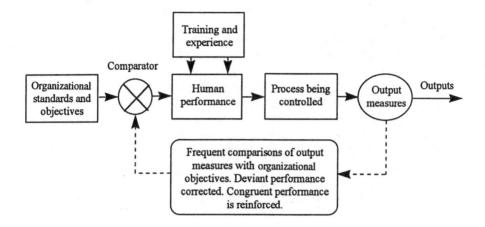

Figure 4.3 Feedback output control requiring frequent comparisons of performance with goals

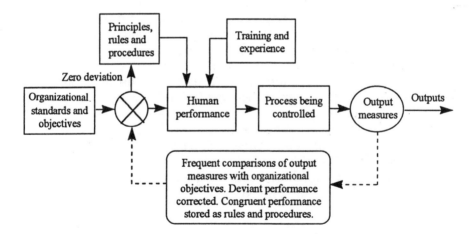

Figure 4.4 Mixed feedback and feedforward controls

It can be seen from Figures 4.2–4.4 that the different administrative control modes share three common features:

- *organizational standards and objectives* that define the system's production and safety goals and also specify the standards of performance required to achieve those goals,

- *human performance,* or the activities carried out by the system's human elements and
- *the process being controlled*—this will obviously vary according to the nature of the technology involved.

In Figure 4.2, human performance—and hence the process being controlled—is strictly standardized by rules and procedures. In the language of the control theorists, these are feedforward devices, with the behavioural means to achieve organizational goals being specified in advance and delivered through rules and procedures, regardless of the local conditions. In the realm of safety-related actions, however, there is an intermittent feedback loop. As discussed in Chapter 3, the response of many bureaucratic organizations to their occasional accidents and incidents is to 'write another procedure' prohibiting those actions implicated in the last event. The result is that the body of procedures increases in an additive fashion as shown in Figure 4.2.

The feedback output control shown in Figure 4.3 defines the opposite end of the prescriptive–discretionary continuum. Here, it is presumed that the workforce understands the organizational goals and possesses the internalized knowledge, skills and abilities to achieve them, using a variety of means. Output measures are frequently compared with organizational objectives. Deviant performance is adjusted to reduce the discrepancy and congruent performance is reinforced.

Within any given technology or system, the balance between feedforward and feedback control modes will depend on several factors:

- the stage reached in the organization's life history,
- the type of work or activities,
- an individual's position within the hierarchy, and
- the amount of training given to individuals.

The Stage Reached in the Organization's Life History

In the early stages of a technology or manufacturing process, work will probably be governed largely by trial and error, or by feedback-driven controls. During this initial period, however, managers and supervisors will be noting those occasions when actions and conditions yield precisely the desired results. In other words, they will be looking out for zero deviations between output measures and organizational goals. When these are observed often enough, it is probable—assuming that the process is reasonably predictable—that they will become expressed as procedures. This will give rise to mixed controls of the kind summarized in Figure 4.4.

As organizational learning proceeds and the work becomes increasingly standardized, so feedforward controls will come to predominate over feedback controls. It is, after all, much cheaper to have the work performed by a relatively untrained labour force controlled by procedures than it is to rely upon the judgement of experienced and highly trained individuals.

In certain professions, the transition from feedback to feedforward may take a very long time. In medicine, for example, doctors have been trained for many centuries to function on a largely discretionary basis. Each patient's illness was thought to pose a unique set of clinical problems, requiring diagnosis and treatment to be determined by a combination of clinical experience and knowledge-based principles. For a variety of political, economic, legal and managerial reasons, however, this traditional approach is now changing to a system in which healthcare is increasingly being delivered on the basis of condition-specific procedures. It is curious that such a bastion of discretionary action as medicine should be moving towards a feedforward mode of control when many other hitherto rule-dominated domains—notably railways and oil exploration and production—are shifting towards performance-based controls and away from prescriptive ones.

Type of Activity

Regardless of the stage of system development, the balance between process (feedforward) controls and output (feedback) controls will be largely shaped by the nature of the tasks that the organization undertakes. Certain activities readily lend themselves to being proceduralized; others do not. Charles Perrow identified two aspects of organizational activity that determine the degree to which it can be pre-programmed.[3]

- The number of 'exceptional cases': That is, the degree to which surprises, novel situations and unexpected events are likely to arise during the course of the work.
- The nature of the 'search process' required to deal with problems. Some problems may be easily resolved by analysis and the application of rule-based solutions; others are little understood and require extensive knowledge-based processing.

Figure 4.5 fleshes out the characteristics of these different kinds of organizational activity and Figure 4.6 gives examples of each of the four task types.

Number of exceptional cases (i.e., new events, situations and problems)

		Few	Many
Nature of search for problem solutions	Easy	*Tasks routine, repetitive, well-structured and predictable. Pre-programmed process control by rules and procedures possible.*	*Tasks non-routine, but the many exceptional cases are relatively simple to analyse. Requires mixture of prescriptive and discretionary performance control.*
	Hard	*Work routine, but problems are sometimes vague and poorly conceptualized. Requires a a mix of prescriptive control by rules and procedures and discretionary performance by the individual.*	*Tasks non-routine, poorly structured and unpredictable. Rules and procedures not applicable. Task performance at the discretion of the individual.*

Figure 4.5   The varieties of organizational activity (after Perrow)

Number of exceptional cases (i.e., new events, situations and problems)

		Few	Many
Nature of search for problem solutions	Easy	Production lines Railways Postal services Construction Traditional banking Road haulage, etc.	Architecture Scientific research Project management Maintenance & repair Oil exploration Police work, etc.
	Hard	Nuclear power plants Chemical process plants Modern aircraft Advanced manufacturing R & D organizations Anaesthesia Geriatric medicine, etc.	Modern military operations Investment banking Macro-economics Politics Recovering beyond design basis accidents Crisis management, etc.

Figure 4.6   Some examples of the various types of activity

Level within the Organization

The higher an individual's position within the organization, the harder it is to proceduralize the job. The task of a train driver, for example, is to start, run and stop the train according to the dictates of the track signalling system and the timetable. How this should be done is clearly prescribed by the railway company's procedures. But there are no procedures setting out how the senior management team should achieve the company's strategic goals. In relatively bureaucratic organizations, therefore, one would expect top management to operate mainly by feedback-driven output controls, with the reverse being true for those at the 'sharp end'. Middle management and first-line supervisors would occupy intermediate positions along the prescriptive–discretionary continuum.

The wide variation in the degree to which individual jobs can be proceduralized has implications not only for the type of controls that are applicable but also for the ease with which errors and violations may be identified or even defined. If a train driver passes a signal at danger, he or she has committed either an unwitting error or a deliberate violation. Both errors and violations are defined in terms of a deviation of action from some intended or required standard of performance. At the 'sharp end' of any industry, there is usually little doubt about the yardsticks from which such departures are gauged. But who is to say whether a top-level strategic decision is correct or not? Such matters can usually only be settled long after the fact, and even then the question of whether or not a given decision was wrong is often a matter of debate.

The Trade-off Between Training and Procedures

The longer and more intensive an individual's training, the less likely is that person to be governed by rigid feedforward controls, and conversely. There are, of course, exceptions. Airline pilots receive extensive training, both initially and throughout their flying careers, yet their activities are highly proceduralized and regulated. The same is true of most branches of the professional military.

Military cultures differ in the autonomy granted to junior commanders and non-commissioned officers. The fighting quality of the German Army—from the Franco-Prussian War in 1870, through the First World War, to the defence of Germany in 1945—was legendary. German ground troops inflicted half as many more casualties than they received from the opposing Allied forces throughout the Second World War, even though the latter had, in later years, massive air superiority over the battlefield. Against the Russians the

Wehrmacht inflicted, on average, six times as many casualties as they received.[4]

Many authorities have speculated on the basis of this military superiority. Was it due to the courage, quality and tenacity of individual German soldiers, or did it stem from the character of their military institutions? While not denying the former possibility, most writers favour the organizational hypothesis.

It has been convincingly argued that Germany created a more effective military organization than was achieved by her enemies. This effectiveness had many facets, but one is of particular relevance here. In its training of junior leaders, particularly NCOs, the Wehrmacht followed the longstanding principle of *Auftragssystem* (literally 'mission system'). Its essence was that a subordinate commander should be able to pursue the tactical goals of his superiors with or without orders. This presupposed a considerable degree of initiative and tactical understanding on the part of these junior leaders. As Alistair Horne observed:

> Far from being a horde of rigid and inflexible robots (which was always one of the most insidious of the Allied misconceptions about the Germans), the Wehrmacht thus had a far greater ability to react or to regain the initiative—especially in a moment of reverse—than was possessed particularly by the British Army of 1944.[5]

If one substitutes the title 'supervisor' for 'NCO' and considers safety rather than war, it is interesting to see whether the 'mission system' idea can be used to improve the guidance of safe behaviour in hazardous technologies. We will return to this issue of decentralization in Chapter 9.

So far, we have been looking at procedures from an organizational perspective. Now we need to take a more psychological stance and consider some of the ways in which rule-related actions can be distinguished. We begin by identifying three levels of human performance.

Three Levels of Performance

Figure 4.7 summarizes the main distinctions between three levels of human performance: the skill-based (SB), rule-based (RB) and knowledge-based (KB) levels. These performance levels, first introduced by Jens Rasmussen,[6] are distinguished by both psychological and situational variables which, together, define an 'activity space' on to which the three performance levels can be mapped.

Human beings control their actions through various combinations of two control modes—the conscious and the automatic. The con-

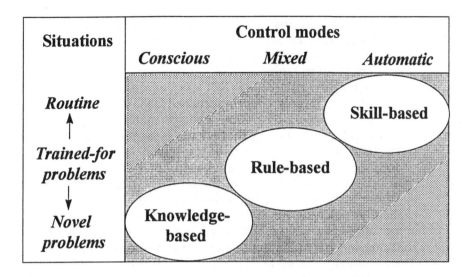

Situations	Control modes		
	Conscious	*Mixed*	*Automatic*
Routine ↑ ┊ *Trained-for problems* ┊ ↓ *Novel problems*	Knowledge-based	Rule-based	Skill-based

Figure 4.7   Location of the three performance levels within an 'activity space' defined by the dominant mode of action control and the nature of the local situation

scious mode is restricted in capacity, slow, sequential, laborious, error-prone, but potentially very smart. This is the mode we use for 'paying attention' to something. But attention is a limited resource, if it is focused upon one thing, it is necessarily withdrawn from other things.

The automatic mode of control is the opposite in all respects. It is largely unconscious—we may be aware of the product (a word, an action, an idea, a perception), but not of the process that created it. The automatic mode is virtually limitless in its capacity—at least, no psychologist has yet discovered these limits. It is very fast and operates in parallel—that is, it does many things at once rather than one thing after another. It is effortless and essential for handling the recurrences of everyday life. But it is a highly specialized community of knowledge structures. It knows only what it knows; it is not a general problem solver like consciousness. Naturally, we prefer to operate in the automatic mode whenever possible.

The second dimension defining the activity space is the nature of the immediate situation. The two extremes of this dimension are highly familiar everyday situations and entirely novel problems, with trained-for problems—or ones for which procedures have been written—lying in between.

The three performance levels can be summarized as follows:

- At the skill-based (SB) level, we carry out routine, highly-practised tasks in a largely automatic fashion with occasional conscious checks on progress. This is what people are very good at most of the time.
- We switch to the rule-based (RB) level when we notice a need to modify our largely preprogrammed behaviour because we have to take account of some change in the situation. This problem is likely to be one that we have encountered before, or have been trained to deal with, or which is covered by the procedures. It is called the rule-based level because we apply memorized or written rules of the kind—*if* (this situation) *then do* (these actions). In applying these rules, we operate by automatically matching the signs and symptoms of the problem to some stored knowledge structure. We may then use conscious thinking to verify whether or not this solution is appropriate.
- The knowledge-based (KB) level is something we come to very reluctantly. Only when we have repeatedly failed to find some pre-existing solution do we resort to the slow and effortful business of thinking things through on the spot. Given time and a forgiving environment to indulge in trial-and-error learning, we often come up with good solutions. But people are not usually at their best in an emergency—though there are some notable exceptions. Quite often, our understanding of the problem is patchy, inaccurate or both. Furthermore, consciousness is very limited in its capacity to hold information; it can store no more than two or three distinct items at a time. It tends to behave like a leaky sieve, allowing things to be lost as we turn our attention from one aspect of the problem to another. In addition, we can be plain scared, and fear like other strong emotions has a way of replacing reasoned action with 'knee jerk' or over-learned responses. This is represented in Figure 4.7 by the bottom right-hand corner of the activity space.

It must be emphasized that these performance levels are not mutually exclusive. All three can coexist at the same time. Consider the process of driving a car. Control of the speed and direction of the car is carried out at the SB level. Negotiating other drivers, pedestrians and the rules of the road occurs at the RB level. However, at the same time that both of these levels are in play, we could also be brooding about some new and difficult life problem at the KB level. Understanding these performance levels helps us to classify the varieties of errors and violations, as discussed below.

Errors and Successful Actions

Human error can be defined as *the failure of planned actions to achieve their desired ends—without the intervention of some unforeseeable event.*[7] The rider is important because it separates controllable action from either good or bad luck. An individual who is struck down by a piece of returning space debris will probably not achieve his or her immediate goal, but this could hardly be termed an error. Similarly, someone who slices a golf ball so that it hits a tree and then bounces fortuitously on to the green may achieve their purpose, but their actions are still erroneous.

There are three elements to this definition: a plan or intention that incorporates both the goal and the means to achieve it, a sequence of actions initiated by that plan, and the extent to which these actions are successful in achieving their purpose. Logically, actions may fail to achieve their goal for one or other of the following reasons:

- The plan is adequate, but the actions fail to go as planned. These are unintended failures of execution and are commonly termed *slips, lapses, trips or fumbles.* Slips relate to observable actions and are commonly associated with attentional or perceptual failures. Lapses are more internal events and generally involve failures of memory.
- The actions may conform exactly to the plan, but the plan is inadequate to achieve its intended outcome. Here, the failure lies at a higher level—with the mental processes involved in assessing the available information, planning, formulating intentions, and judging the likely consequences of the planned actions. These errors are termed *mistakes*, and have been further divided into two subcategories: *rule-based mistakes* and *knowledge-based mistakes.* Rule-based mistakes involve either the misapplication of normally good rules, the application of bad rules, or the failure to apply a good rule (a violation). Knowledge-based mistakes occur when we have run out of prepackaged solutions and have to think out problem solutions on line. This, as discussed above, is a highly error-prone business. The various error types are summarized in Figure 4.8.

As mentioned earlier, all errors involve some kind of deviation. In the case of slips, lapses, trips and fumbles, actions deviate from the current intention. In the case of mistakes, however, the departure is from some adequate path towards the desired goal.

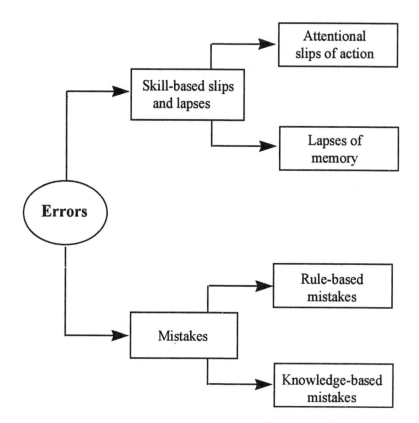

Figure 4.8   Summary of the principal error types

Violations and Compliant Action

Violations are deviations from safe operating procedures, standards or rules.[8] Such deviations can be either deliberate or erroneous (for example, speeding without being aware of either the speed or the local restriction). However, we are mostly interested in deliberate violations, where the actions—though not their possible bad consequences—are intended. These non-malevolent acts should be distinguished from sabotage in which both the act and the damaging outcome are intended.

Three major categories of safety violation have been distinguished: routine, optimizing and necessary violations. In each case, the decision not to abide by safe operating procedures is shaped by both organizational and individual factors, though the balance of these influences varies from one type of violation to another.

Routine violations typically involve corner-cutting at the skill-based level of performance—taking the path of least effort between two task-related points. These short-cuts can become an habitual part of the person's behavioural repertoire, particularly when the work environment is one that rarely sanctions violations or rewards compliance. Routine violations are also promoted by 'clumsy' procedures that direct actions along what seems to be a longer-than-necessary pathway.

Optimizing violations—or violating for the thrill of it—reflect the fact that human actions serve a variety of motivational goals and that some of these are quite unrelated to the functional aspects of the task. Thus, a driver's functional goal is get from A to B, but in the process he or she can optimize the joy of speed or indulge aggressive instincts. Tendencies to optimize non-functional goals can become a part of an individual's performance style. They are also characteristic of particular demographic groups such as young male drivers.

Whereas routine and optimizing violations are clearly linked to the attainment of personal goals (that is, least effort and thrills), necessary violations have their primary origins in particular work situations. Here, non-compliance is seen as essential in order to get the job done. Necessary violations are commonly provoked by organizational failings with regard to the site, tools or equipment—see the shunting example given in Chapter 3. In addition, they can also provide an easier way of working. The combined effect of these two factors often leads to these violations becoming routine rather than exceptional.

Correct and Incorrect Actions

Here, the terms 'correct' and 'incorrect' relate to the accuracy of risk perception. A *correct action* is one taken on the basis of an accurate risk appraisal. An *incorrect action* is one based upon an inaccurate or inappropriate assessment of the associated risks.

Successful actions—with respect to personal goals—need not necessarily be correct actions—with regard to the accuracy of the risk assessment. As indicated earlier, driving at 100 mph could achieve the personal goals of arriving sooner and of experiencing the thrill of speed, but it is clearly not 'correct' given the increased risks associated with a relatively unknown physical regime and its reduced error margins. Similarly, compliance is not automatically 'correct', nor a violation incorrect. It all depends on the local conditions and the adequacy of the procedures.

What is 'correct' is often unknowable in advance, and frequently unknown at the time. In principle, however, it could be expressed in probabilistic terms. For example, the likelihood of being involved in a fatal accident while travelling within the speed limit on a British motorway can be assessed from data relating accident frequencies to miles of exposure.[9] Similarly, the extent to which this risk is increased through exceeding the speed limit by varying amounts can also, in principle, be determined.

The Quality of the Available Procedures

So far, we have looked at rule-related behaviours from both an organizational and a psychological perspective. Now we will adopt a control theory perspective and consider the effectiveness of safety procedures. There are three possibilities: good rules, bad rules or non-existent rules—no rules. Here, 'good' entails both correctness and the furtherance of acceptable personal goals (that is, efficiency, least effort). Bad rules are either inappropriate for the particular circumstances, or incorrect, or both.

Unlike the procedures governing productive activities, those written to ensure safe working suffer from a lack of requisite variety. In virtually all hazardous operations, the variety of possible unsafe behaviours is very much greater than the variety of required productive behaviours. Thus, while it may be appropriate to control productive performance through feedforward prescriptions (see again Figure 4.2), the requisite variety of the procedures necessary to govern safe behaviour will always be less than the possible variety of unsafe situations. As the control theorist, Ross Ashby, put it, 'only variety can destroy variety'. In other words, only variety in the controlling agency can reduce variety in the to-be-controlled outcome.[10]

We can illustrate this problem with a simple domestic example. The procedures necessary to guide the preparation of, say, minestrone soup can be conveyed in a few sentences. But the procedures necessary to guarantee that this task will be performed with absolute safety could fill several books, and even then it is unlikely that all the possible hazards and accident scenarios would have been covered. Of course, we can anticipate some of the probable hazards (harmful bacteria, sharp knives, scalding, electrocution, inflammable and poisonous gases and so on) and seek to regulate our cook's behaviour with regard to these obvious dangers. And this is what organizations try to do in relation to specific operational hazards. But there is no way in which all the possible combinations of dangers and their related accident scenarios could be anticipated. In essence, therefore, wholly safe behaviour can never be controlled entirely by feedforward prescriptions. There will be always be situations that are either not covered by the rules or in which they are locally inapplicable. In other words, there will always be 'bad rule' situations and 'no rule' situations.

This does not mean, of course, that organizations should give up the attempt to formulate safety rules. Not only are such rules important for guiding safe behaviour in relation to identified and understood hazards, they also constitute an important record of the organization's learning about its operational dangers. Since people change faster than jobs, such a record is crucial for the dissemination of safety knowledge throughout the system. But the requisite variety problem means that this collection of safe operating procedures will never be wholly comprehensive nor universally applicable.

Six Kinds of Rule-related Behaviour

On the basis of the various distinctions discussed above, we can now identify six types of rule-related behaviour, as shown in Figure 4.9. Here, there are three kinds of rule-related situations and two kinds of performance. In the latter case, 'correct performance' refers to actions that are both correct (in regard to risk assessment) and successful (in achieving personal goals). 'Erroneous performance', on the other hand, refers to actions that are both incorrect and unsuccessful.

	Good rules	**Bad rules**	**No rules**
Correct performance	Correct compliance	Correct violation	Correct improvization
Erroneous performance	Misvention	Mispliance	Mistake

Figure 4.9   Six varieties of rule-related performance

The main characteristics of these six behaviours are summarized below, and real-life examples of these actions are given in the next section.

- *Correct compliance:* correct (and safe) performance achieved through adhering to appropriate safety rules.
- *Correct violation:* correct performance achieved by deviating from inappropriate rules or procedures.

- *Correct improvisation:* a course of action taken in the absence of appropriate procedures that leads to a safe outcome.
- *Misvention:* behaviour that involves both a deviation from appropriate safety rules and error(s), leading to an unsafe outcome.
- *Mispliance:* behaviour that involves mistaken compliance with inappropriate or inaccurate operating procedures, leading to an unsafe outcome.
- *Mistake:* an unsafe outcome resulting from an unsuitable plan of action carried out in the absence of appropriate procedures (that is, a knowledge-based mistake).

There is one category of unsafe actions that is not captured by the logic of Figure 4.9, and that is the incorrect, but successful, violation of good rules.[11] Here, personal goals are achieved as planned despite faulty hazard assessments. These successful violations—not the same as correct violations—are the breed stock from which occasional misventions emerge. Successful violations, though often inconsequential in themselves, create the conditions that promote dangerous misventions. These will include overconfidence in personal skills and a chronic underestimation of the hazards.

What are the relative frequencies with which these various types of behaviour are likely to occur? Since safety rules usually evolve by a process of natural selection, we can reasonably expect that good rules will outnumber bad rules, and that 'no rule' situations will be rare. It is also highly probable that compliance will occur more frequently than non-compliance. Thus, correct compliance is likely to be the commonest mode of behaviour with correct violations, mispliances and improvizations—either successful or unsuccessful—as the least frequent types. The evidence from field studies indicates that successful, but incorrect, violations are relatively commonplace, although they number far fewer than correct compliances. Misventions would comprise a much smaller subset of successful but incorrect violations.

Some Real-life Examples

Misventions

Chernobyl was an accident initiated almost entirely by misventions.[12] Of the seven well-documented unsafe acts leading up the explosion of the Ukrainian RBMK nuclear reactor in April 1987, six were misventions of one kind or another. The first (and perhaps most critical) of these was the operators' mistaken decision to continue the testing of the voltage generator even though an initial operating error had caused the power level to fall to 7 per cent full power. The characteristics of the RBMK reactor were such that operations at these low-power regimes created a positive void coefficient in the reactor's core. This can lead to runaway reactivity. The station operating procedures strictly prohibited any operations below 20 per cent full power (the initial intention was to run the experiment at 25 per cent full power). Subsequently, operators successively switched off various engineered safety systems in order to complete the experiment, and in so doing made the subsequent explosions inevitable. Here was a sad and remarkable case in which a group of highly motivated, prize-winning operators managed to destroy an elderly, but relatively well defended, reactor without the assistance of any technical failures.

Successful Compliance and Mistakes

Successful compliance is generally not recorded since it is rarely implicated in an accident. Recently, however, the US Nuclear Regulatory Commission sponsored a human factors inspection programme into the circumstances surrounding a number of potentially significant operational events in US nuclear power plants.[13] Of the 21 inspection reports, approximately half were designated as successful recoveries and the remainder as less successful recoveries. All events ended safely, but in 11 cases, the operators showed confusion or uncertainty as to the causal conditions and made errors in their attempts to restore the plant to a safe state. In contrast, the actions of the successful crews were exemplary and, on occasions, inspired. An important discriminator between the successful and less successful recoveries was the state of the plant when the off-normal conditions arose.

In most of the successful recoveries, the plants were operating at or near full power. This is the 'typical' condition for a nuclear power plant—pushing out megawatts to the grid. It is for this condition that most of the emergency operating procedures are written. Nearly all successful recoveries were facilitated by these procedures and their associated training. The less successful recoveries, on the other hand, occurred mostly during low-power or shutdown conditions. These are perfectly normal plant states—necessary for maintenance, repairs and refuelling—but they constitute a relatively small part of the total plant life history. Few procedures are available to cover emergencies arising in these atypical states. As a result, the operators had to 'wing it' and, on several occasions, made errors that exacerbated the original emergency and delayed recovery. The crews operated outside the rules because no rules were provided for handling emergencies in these still dangerous plant conditions. As a result, they were forced to improvise in situations they did not fully understand. While not ultimately unsuccessful (the plants were eventually restored to a safe condition), many of their actions along the way fitted the mistake category, identified in Figure 4.9.

Mispliances and Successful Violations

The *Piper Alpha* disaster on 6 July 1988 provided examples of both mispliances and successful violations. The emergency procedures required platform personnel to muster in the galley area of the accommodation. Ed Punchard describes their behaviour:

> All over the rig, people were instinctively following their training and emergency instructions. In the absence of any form of announcement, most were trying to make their way to the galley to muster, have a head count, and take instructions. After all, that was what they were trained to do.[14]

Tragically, the accommodation was directly in the line of the fireball that erupted at 11.22 pm. Most of those who complied with the emergency procedures were burned or suffocated to death.

At the time when the first explosion occurred (around 9.58 pm), Ed Punchard, the diving superintendent, was in a diving office on one of the lower decks of the platform with some of his diver colleagues. Their first action was run up the stairs to the 85-foot level, hoping to reach the lifeboat on the 107-foot level. On the way, they encountered thick smoke and heard the sound of escaping gas, so they started to descend and, with the help of a rope, reached the spider-deck on the north-west leg. From there, Punchard climbed down a steel ladder and jumped on to a rescue boat. Whether the manner of his escape was an act of deliberate violation or sheer necessity is not clear. What is evident, however, is that those people who deviated from the mustering instructions formed the majority of the survivors.

Military history is rich in both mispliances and successful violations, though only the latter tend to be celebrated, such as Nelson's violation at Copenhagen when he chose to scan the flagged order to disengage from the enemy holding the telescope to his blind eye. Of course, had he subsequently gone on to lose the battle, history would have viewed such a misvention in a different light. Successful violations are often taken as the mark of the great commander, while mispliances are seen as characteristic of inferior ones. Sun Tzu, in his 2500-year-old treatise on 'The Art of War', recommended judicious violations:

> If fighting is sure to result in victory, then you must fight, even though the ruler forbid it; if the fighting will not result in victory, then you must not fight even at the ruler's bidding.[15]

Successful Improvization

One of the most dramatic examples of successful improvization occurred on 19 July 1989 aboard United 232, a McDonnell Douglas DC-10, commanded by Captain Alfred C. Haynes. While flying at cruise altitude, the fan rotor of the tail-mounted number two engine disintegrated and cut through all three of the aircraft's hydraulic systems. The probability of losing all three hydraulic systems was considered by the designers to be less than one-in-a-billion (10^{-9}) and no emergency procedures were available to cover this almost unthinkable possibility. Al Haynes described their plight as follows:

> That left us at 37 000 feet with no ailerons to control roll, no rudders to co-ordinate a turn, no elevators to control pitch, no leading edge devices to help us slow down, no trailing edge flaps to be used in landing, no spoilers on the wings to slow us down in flight or help braking on the ground, no nosewheel steering and no brakes. That did not leave us a great deal to work with.[16]

Forty-five minutes later, the aircraft crash-landed at Sioux City, Iowa. Of the 285 passengers and 11 crew members on board, 174 passengers and 10 crew survived. About 14 seconds into the emergency, the aircraft had reached 38 degrees of bank to the right, on its way to rolling over on its back and then falling out of the sky. At this point, Al Haynes took control of the aircraft, slammed the number one (left engine) throttle closed and opened the number three (right engine) throttle all the way, with the result that the right wing started to come back up. After that, the two pilots—plus a third flying as a passenger—learned to use the differential thrusts of the two remaining wing-mounted engines to skid the aircraft left or right in descending turns. In this way, they coaxed it into Sioux City and might have landed safely had the aircraft not been affected by uncontrollable manoeuvres (phugoids) while low on the approach. To save the lives of 184 out of a total of 296 people aboard—all of whom would have almost certainly died in an uncontrollable crash—was a remarkable feat of airmanship and crew resource management.

Assembling the Big Picture

This chapter has examined some of the behavioural options available to people working in complex technologies. Since all such behaviour is governed by administrative controls, we began by looking at the principal means used by systems to limit people to safe and productive courses of action. These were represented along a dimension

ranging from externalized feedforward procedures, that rigidly prescribe the allowable actions, to largely discretionary feedback controls,
based upon internalized knowledge, skills and experience. Several
factors influenced which of these two modes would predominate:
the stage reached in the life history of the system; the type of operations carried out by the system; an individual's level within the
organization; and the amount of training deemed necessary.

The next part of the chapter discussed a number of ways in which
human performance could be segmented. Three levels of performance were identified—the skill-based (SB), rule-based (RB) and
knowledge-based (KB) levels—each defined by the interaction between the modes of action control (conscious or automatic) and the
nature of the immediate situation (routine or problematic). Three
major error types, or unsuccessful actions, were linked to these performance levels: skill-based slips and lapses, rule-based mistakes
and knowledge-based mistakes. We then identified three ways in
which people can violate procedures: routine violations, optimizing
violations and necessary violations. We also distinguished between
correct and incorrect actions on the basis of whether or not they
involved an appropriate assessment of the dangers involved in a
planned course of action.

Next, the point was made that safety procedures could never anticipate all the ways in which harm could come to people in the
course of their work. They lacked the requisite variety to match these
many dangers. As a result, work situations would fall into one of
three groups: those covered by good procedures, those for which the
available procedures were inappropriate or ill-matched, and those
for which no procedures had been written. These three conditions—
good rules, bad rules and no rules—were then combined with correct
and erroneous performance to generate six kinds of rule-related behaviour: correct compliance, correct violation, correct improvization,
misvention (mistaken circumvention), mispliance (mistaken compliance) and mistakes (failures of improvization).

Both organizations and human behaviour are complex, so it is
hardly surprising that this chapter has invoked a bewildering number
of distinctions. Now the time has come to put them, quite literally,
into one big picture. This is shown in Figure 4.10 which offers two
principal routes for action: one that goes directly to correct and successful performance; and the other showing a variety of pathways to
what are mostly unsafe behaviours. In both cases the starting point is
the question: was the task (or situation) covered by procedures or
training? Human actions can only be understood within a given
context which, in this case, is a work-related task or situation. Although most activities within well established systems will have
been anticipated in one way or another, sometimes totally novel

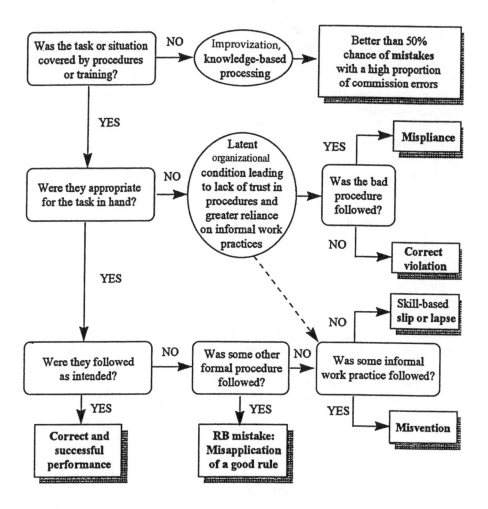

Figure 4.10 Summarizing the varieties of rule-related behaviours

situations arise in which people have to improvize a suitable course of action on the basis of knowledge-based processing. When the individuals concerned are both highly skilled and highly experienced (like Captain Al Haynes), there seems to be a 50:50 chance of coming up with the right answers. Mostly, however, the odds are much lower.

While procedures may have been written for a particular task, they may not necessarily be appropriate for, or applicable to, the local situation. In the nuclear power industry, for example, some 60 per cent of all human performance problems are associated with bad procedures. Wrong or inapplicable procedures are latent conditions that will cause the workforce to lose trust in the official procedures. They may either ignore them altogether or follow unofficial procedures of their devising—known in some industries as 'black books'. People faced with a bad procedure have two options: they can either follow the procedure (a mistaken compliance or mispliance) or not (a correct violation).

The next level of Figure 4.10 poses the question: were the good procedures followed as intended? If the answer is 'yes', we have then completed all the stages necessary for correct and successful performance. As we have seen, there are a number of ways in which human actions can deviate from this desirable pathway. We can commit a rule-based mistake. For example, we can misread the situation, fail to recognize it as an exception, and apply a rule or procedure that is appropriate for most situations of this general kind but not this particular one. Second, we could follow some informal working practice and commit a mistaken violation (misvention). Finally, we could make a slip or suffer memory lapse so that the actions, although intended to follow the procedure, do not turn out as planned.

We have now covered most of the human behaviours implicated in organizational accidents. In the next chapter, we will focus on a single type of error—maintenance omissions during installation or reassembly. As we shall see, the frequency with which this one error type crops up as a contributor to organizational accidents more than justifies such close attention.

Notes

1   P. Johnson and J. Gill, *Management Control and Organizational Behaviour*, (London: Paul Chapman Publishing, 1993).
2   J. Reason, D. Parker, R. Lawton and C. Pollock, *Organizational Controls and the Varieties of Rule-Related Behaviour*, paper presented to the UK Economic and Social Research Council's Meeting on Risk in Organizational Settings, London, May 1995.
3   C. Perrow, 'A framework for the comparative analysis of organizations', *American Sociological Review*, **32**, 1967, pp. 194–208.
4   A. Horne, *The Lonely Leader: Monty 1944–45*, (London: Macmillan, 1994).
5   Ibid., pp. 191–2.
6   J. Rasmussen, 'Human errors: a taxonomy for describing human malfunction in industrial installations', *Journal of Occupational Accidents*, **4**, 1982, pp. 311–33. See also J. Rasmussen, Skills, rules, knowledge: signals, signs and symbols and other distinctions in human performance models. *IEEE Transactions: Systems, Man and Cybernetics*, SMC-13, 1983, pp. 257–67.
7   J. Reason, *Human Error*, (New York: Cambridge University Press, 1990).

8 See J. Reason, D. Parker and R. Free, *Bending the Rules: The Varieties, Origins and Management of Safety Violations*, (Leiden: Faculty of Social Sciences, University of Leiden, 1994).

9 D.F. Finch, P. Kompfner, C.R. Lockwood and G. Maycock, *Speed, Speed Limits and Accidents*, Project Report No. 58, (Crowthorne: Transport Research Laboratory, 1994).

10 W.R. Ashby, *An Introduction to Cybernetics*, (London: Chapman & Hall, 1956).

11 Tony Muschara of the Institute of Nuclear Power Operations in Atlanta has suggested a further addition to this list – *malpliance*, malicious compliance with procedures that are manifestly inapplicable or wrong.

12 J. Reason, 'The Chernobyl errors', *Bulletin of the British Psychological Society*, **40**, 1987, pp. 2201–6.

13 NUREG/CR-6093, *An Analysis of Operational Experience during Low Power and Shut Down and a Plan for Addressing Human Reliability Assessment Issues*, (Washington, DC: US Regulatory Commission, 1994).

14 E. Punchard, *Piper Alpha: A Survivor's Story*, (London: W.H. Allen & Co., 1989), p. 128.

15 L. Giles, *Sun Tzu on the Art of War*, (Singapore: Graham Brash, 1988 (first published in 1910)).

16 A.C. Haynes, 'United 232: Coping with the loss of all flight controls', *Flight Deck*, **3**, 1992, pp. 5–21, see p. 7.

5 Maintenance can Seriously Damage your System

Close Encounters of a Risky Kind

In Chapter 3 we discussed some of the problems created by automated systems in modern technologies. Distancing people from the processes they nominally control reduces the amount of 'hands on' contact, and hence the occurrence of slips, lapses, trips and fumbles, but it also increases the likelihood of certain kinds of mistakes— notably mode confusions—that may have more disastrous consequences than those the designers were seeking to avoid.

In this chapter, we look at the opposite end of this 'hands on' spectrum, at maintenance-related activities in which the amount of direct contact between people and the system has certainly not diminished; indeed, in some areas it has actually increased. The term 'maintenance-related' includes unscheduled repairs, inspections, planned preventive maintenance, calibration and testing.

It will be argued that these close encounters with the technical components of a system probably constitute the single largest human performance problem facing most hazardous technologies. By the same token, they also offer the greatest opportunity for human factors improvements. Errors by pilots, control room operators and other 'sharp-end' personnel may add the finishing touches to an organizational accident, but it is often latent conditions created by maintenance lapses that either set the accident sequence in motion or thwart its recovery.

The chapter will present evidence to support the following assertions:

- There have been a large number of organizational accidents in which maintenance failures were either a significant cause or an exacerbating feature.

- Of the three kinds of human activity that are universal in hazardous technologies—control under normal conditions, control under emergency conditions and maintenance-related activities—the latter poses the largest human factors problem.
- Of the two main elements of maintenance—the disassembly of components and their subsequent reassembly—the latter attracts by far the largest number of errors.
- Of the various possible error types associated with the reassembly, installation or restoration of components, omissions—the failure to carry out necessary steps in the task—comprise the largest single error type.

We will focus primarily on aviation and nuclear power generation because these two areas offer the best documented data on maintenance-related failures. But they are not special cases. There is every reason to suppose that similar patterns of occurrence are present in all hazardous technologies. Regardless of the domain, maintenance-related activities share a large number of common characteristics, particularly the taking apart and the putting back together of components. In addition, of course, they all involve close encounters with the human hand. And it is this frequency of contact, and hence greater error opportunity, rather than any lack of competence on the part of maintenance personnel that lies at the heart of the problem.

The chapter then goes on to examine some of the ways in which these commonly occurring failures can be managed. A more fundamental remedy, however, is for designers and manufacturers to recognize that the risk of a 'weak' component failing due to lack of maintenance may now be considerably less than the risk of a 'healthy' component being damaged or omitted during maintenance.

Organizational Accidents and Maintenance Failures

It is neither possible nor necessary to provide a comprehensive list of all the organizational accidents in which maintenance-related failures have been implicated. For our purposes, it is sufficient to present some well documented examples from a range of hazardous technologies.

Apollo 13

Shortly after the launch of Apollo 13 on 11 April 1970, an explosion occurred within the Service Module, blowing out its side.[1] The explosion originated within a tank of liquid oxygen and had been initiated by the first activation by the crew of a set of rotating paddles designed to stir the liquid oxygen at intervals throughout the mission. Just prior to the launch, the paddles had been bench-tested. But they had been connected to the wrong power supply. The effect of this error was to burn off the insulating material associated with the electrical supply to the paddles. This, in turn, converted the oxygen cylinder into a bomb waiting to be set off as soon as the paddles were switched on, at which point a spark ignited the liquid oxygen. The rest, as they say, is history. The spacecraft circled the Moon, with the crew—Astronauts Lovell, Haise and Swigert—using the life-support system of the Lunar Module during the extremely perilous re-entry into the Earth's atmosphere. This was one of many events in which the skills of the flight crew, acting as a last line of defence, saved the day following a maintenance-initiated emergency.

Flixborough

In 1974, a temporary pipe at the Nypro Factory in Flixborough failed, releasing 50 tons of hot cyclohexane.[2] The cyclohexane mixed with the air and exploded, killing 28 people and destroying the plant. The production process normally used six reactor vessels, connected in series, each joined to the next by a short 28-inch pipe with expanding bellows. Two months before the disaster, one of the reactor vessels— number five in the series—developed a crack and had to be removed. It was replaced by an improvized 20-inch bypass pipe connecting vessels four and six. The bypass pipe was inadequately supported and, because it entered into the existing bellows at either end, it was free to move about or 'squirm' when the pressure rose above normal—causing the cyclohexane to be released into the atmosphere.

Three Mile Island

Two separate maintenance failures were implicated in the Three Mile Island nuclear power plant accident that occurred on 28 March 1979.[3] One initiated the emergency; the other delayed its recovery. The first failure occurred when a maintenance crew was attempting to renew the resin for the treatment of the plant's water. A small volume of water found its way into the plant's instrument air system, tripping the feedwater pumps. This, in turn, cut the water flow to the steam generator, and tripped the turbine, preventing the heat of the primary cooling system from being transferred to the cooler water in the secondary system. At this point, the emergency feedwater pumps came on automatically. However, the pipes from the emergency feedwater storage tanks were blocked by closed valves, erroneously left shut during maintenance two days earlier. With no heat removal, there was a rapid rise in core temperature and pressure which caused the reactor to 'scram'—the control rods were automatically lowered into the core, absorbing neutrons and stopping a chain reaction. However, because decaying radioactive materials still produce heat, the temperature and pressure within the core increased further, causing a pilot-operated relief valve to open. This was supposed to flip open and then close but it remained stuck in the open position. A hole was thus created in the primary cooling system through which radioactive water, under high pressure, poured into the containment area and then down into the basement. It took a further 16 hours to restore the plant to a safe state.

American Flight 191, Chicago O'Hare

On the afternoon of 25 May 1979, the day before the Memorial Day weekend, an American Airlines DC-10 departed for Los Angeles.[4] Just before it rotated into the takeoff attitude, pieces of the port engine fell away from the aircraft. Shortly afterwards, the entire engine and pylon assembly tore itself loose, flew over the wing and then smashed down on to the runway. The DC-10 continued to climb away, but at 300 feet it started to bank sharply to the left. The bank steepened and the nose began to fall. Within seconds, it crashed into the ground and exploded, killing all 271 people aboard. The accident investigators discovered fatigue cracking and fractures on the engine pylon. A crescent-shaped deformation on the upper flange of the pylon bulkhead strongly suggested that the flange had been cracked when the pylon was being removed from the wing or reinstalled during maintenance. The investigation revealed that the port engine and pylon had been removed eight weeks earlier to replace the self-aligning bearings in compliance with a McDonnell Douglas Service Bulletin. As a result of this finding, the Federal Aviation Agency took the unprecedented step of grounding all US-registered DC-10s for further investigation. These inspections revealed that six supposedly serviceable DC-10s had similar fractures in the upper flanges of their pylon rear bulkheads. Four of these aircraft belonged to American Airlines and two to Continental Airlines. It was later discovered that both American and Continental had devised special procedures for carrying out the replacement of the forward and rear bulkheads' self-aligning bearings, as required by the manufacturer's Service Bulletin. Although the manufacturer's Bulletin recommended that the engines be removed before the pylons, both airlines had devised what they believed to be a more efficient technique, namely removing the engine and the pylon as a single unit. The investigators concluded that the engine and pylon separation resulted from damage inflicted by these improper maintenance procedures.

Bhopal

On the night of 2–3 December 1984 a leak of methyl isocyanate from a small pesticide plant devastated the Indian city of Bhopal, killing 2500 people and injuring several thousands more.[5] The immediate cause of the discharge was the influx of water into a methyl isocyanate storage tank. Its presence there involved a tangled story of botched maintenance, improvized bypass pipes, failed defences, drought and flawed decision-making on the part of both management and politicians. Among the contributing maintenance failures were a disconnected flare tower, an inoperable refrigeration plant and a failure to regularly clean pipes and valves.

Japan Air Lines, Flight JL 123, Mount Osutaka

On 12 August 1985, a JAL Boeing 747, on an internal flight, crashed into the side of Mount Osutaka, about 100 km west of Tokyo's Huneda Airport.[6] The crash resulted in the highest death toll ever to occur in a single-aircraft accident. After a lengthy and painstaking investigation, it was established that the principal cause of the accident was a botched fuselage repair carried out more than seven years earlier. In splicing the repaired portion of the pressure bulkhead to the original, part of the splice had been wrongly assembled. Over a small section of the splice, two separate doubler plates, instead of one continuous one, were used as reinforcement. The gap in the doubler plating meant that the splice was joined by only a single row of rivets, instead of two rows. This reduced the assembly's resistance to fatigue by some 70 per cent. These problems were not discovered in subsequent maintenance checks, mainly because the area in which the splicing had been carried out was inaccessible to visual inspections.

Piper Alpha

The *Piper Alpha* disaster, which occurred on the evening of 6 July 1988, resulted in the deaths of 165 of the 226 people on board the North Sea oil and gas platform, together with two crew members of a nearby rescue vessel.[7] The explosion was caused by a leak of condensate which occurred when members of the night shift attempted to restart a pump that had been shut down for maintenance. Unknown to them, a pressure safety valve had been removed from the relief line of the pump and a blank flange assembly, that had been fitted at the site of the valve, was not leak-tight. Their unawareness of the valve removal was the result of communication failures at the shift handover earlier in the evening, together with a breakdown of the permit-to-work system relating to the valve maintenance.

Clapham Junction

At 0810 on Monday, 12 December 1988, a northbound commuter train ran into the back of a stationary train in a cutting just south of Clapham Junction station. A third train, going south, ran into the wreckage of the first train.[8] Thirty-five people died and 500 were injured. The immediate cause of the crash was a signal that failed with a green aspect, concealing the presence of the stationary train from the driver of the northbound commuter train until it was too late. This failure was directly due to the working practices of a technician engaged in rewiring the signal on the previous day. Rather than cutting off or tying back the old wires, the technician merely pushed them aside. It was also his practice to re-use old insulating tape, though on this occasion no tape at all was wrapped around the bare ends of the wire. As a result, the wire came into contact with nearby equipment causing a 'wrongside' signal failure.

Phillips 66 Company

At 1300 on 23 October 1989 an explosion and fire ripped through the Phillips 66 Company Houston Chemical Complex in Pasadena, Texas.[9] Twenty-three workers were killed and more than 130 injured. Property damage amounted to nearly a billion dollars. The accident was caused by the release of extremely flammable gases that occurred during regular maintenance operations on one of the plant's polyethylene reactors. In less than two minutes, the vapour cloud came into contact with an ignition source and exploded with the force of 2.4 tons of TNT. Contract maintenance personnel were engaged in removing 'plugs' of solidified polyethylene from the settling legs. There was a single-ball valve at the point where the legs joined the reactor pipes. These valves were kept open during production, so that the polyethylene particles could settle into the leg, and closed during maintenance operations. On this occasion, the valve was opened in error. An air hose that supplied the air pressure to open or close the valve had been connected the wrong way round (the two ends were identical). This allowed the valve to open even when the actuator switch was in the closed position. The flammable gases escaped through the open valve.

In these brief accounts of maintenance-induced organizational accidents, we have deliberately sacrificed detail for scope. The intention was to convey something of the extent to which maintenance errors have been implicated in a wide range of accidents across different domains. But it must be emphazised that maintenance errors—like human failures in any other sphere—are not just isolated causes; they are themselves the consequences of upstream organizational factors or latent conditions.

Activities and their Relative Likelihood of Performance Problems

There are many ways of classifying human performance and its associated errors, each one having its uses and limitations. In Chapter 4, for example, we identified three main classes of human error: skill-based slips and lapses, rule-based mistakes and knowledge-based mistakes. Such a *causal* taxonomy is helpful in locating the underlying mental processes, but it is difficult for non-specialists to apply. Left to their own devices, engineers and quality inspectors are much more likely to classify human performance problems according to their *consequences* for the system (for example, missing fastenings, improper installation, tools left behind and the like). Although it may obscure the underlying cognitive mechanisms, this kind of information is both widely available and easy to interpret.

One of the advantages of these *consequential* classifications is that there is usually little doubt as to what the person was doing when the error occurred. Such breakdowns of performance problems by activity can be very revealing, as will be shown below.

It is useful, when considering the varieties of human performance, to start with a set of activities that are carried out within all hazardous technologies. A preliminary list is set out below:

- Control under normal conditions
- Control under abnormal or emergency conditions
- Maintenance, calibration and testing.

To this list we should properly add the preparation and application of procedures, documentation, rules, regulations and administrative controls but, as these were considered in Chapter 4, we will limit our attention here to the three activities identified above.

From this list of universal human activities it is possible to make a preliminary assessment of their relative likelihood of yielding human performance problems. To do this, we need to ask three questions.

- *The 'hands on' question.* What activities involve the most direct human contact with the system and thus offer the greatest opportunity for active failures (errors and violations) to have an adverse effect upon the system?
- *The criticality question.* What activities, if performed less than adequately, pose the greatest risks to the safety of the system?
- *The frequency question.* How often are these activities performed in the day-to-day operation of the system as a whole?

It would be reasonable to expect that an activity scoring 'high' on all three of these questions is the one most likely to be associated with

Table 5.1   **The relative likelihood of human performance problems in the universal human activities**

Activity	'Hands on'	Criticality	Frequency
Normal control	Low	Moderate	High
Emergency control	Moderate	High	Low
Maintenance-related	High	High	High

human performance problems. The results of this analysis are summarized in Table 5.1.

Maintenance-related work—scoring 'high' on all three criteria—emerges as the activity most likely to generate human performance problems of one kind or another. To what extent is this prediction borne out by the available evidence?

The most relevant data come from nuclear power operations. Table 5.2 shows a compilation of the results of three surveys: two carried out by the Institute of Nuclear Power Operations (INPO) in Atlanta, and one conducted by the Central Research Institute for the Electrical Power Industry (CRIEPI) in Tokyo. The inputs for the INPO investigation were significant event reports filed by US nuclear utilities. In the first INPO study, 87 significant event reports yielded 182 root causes.[10] In the second INPO study, 387 root causes were identified from 180 significant event reports.[11] The data for the Japanese study came from 104 standardized event reports from the CRIEPI-associated utilities.[12]

These data bear out the earlier analysis. In all three studies more than half of all the identified performance problems were associated with maintenance, testing and calibration activities.

Table 5.2   **A compilation of the results of three studies showing the relationship between activities and performance problems**

Activities	Mean proportions of performance problems (% of total)	Ranges (%)
Maintenance-related	60	55–65
Normal operations	16	8–22
Emergency operations	5	2–8

The Vulnerability of Installation

The next question concerns which aspect of maintenance, testing and calibration is most likely to be associated with less-than-adequate human performance. Regardless of the domain, all maintenance-related activities require the removal of fastenings and the disassembly of components, followed by their reassembly and installation. Thus, a large part of maintenance-related activity falls into the categories of either disassembly or installation.

Once again, there are *a priori* grounds for identifying one of these tasks—installation—as being the most likely to attract human performance problems. The reasons for this are made clearer by reference to Figure 5.1, showing a bolt with eight marked nuts attached to it. This represents the maintenance task in miniature. Here, the requirement is to remove the nuts and to replace them in some predetermined order. For the most part, there is only one way to remove the nuts, with each step being naturally cued by the preceding one. The task is one where all the necessary knowledge is 'in the world' rather than 'in the head'.[13]

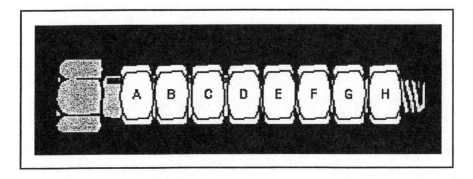

Figure 5.1   The bolt-and-nuts example

In the case of installation, however, there are over 40 000 ways in which the nuts can be reassembled in the wrong order (factorial 8). And this takes no account of any possible omissions. Moreover, the likelihood of error is further compounded by the fact that many possible omissions and misorderings may be concealed during the reassembly process. Thus, the probability of making errors during installation is very much greater than during disassembly, while the chances of detecting and correcting them are very much less.

The available evidence supports the prediction that installation will be especially vulnerable to errors. Listed below are the top seven causes of 276 inflight engine shutdowns (IFSDs) in Boeing aircraft.[14]

- Incomplete installation (33%)
- Damaged on installation (14.5%)
- Improper installation (11%)
- Equipment not installed or missing (11%)
- Foreign object damage (6.5%)
- Improper fault isolation, inspection, test (6%)
- Equipment not activated or deactivated (4%).

These data show that various forms of faulty installation were the top four most frequent causal categories, together comprising over 70 per cent of all contributing factors. Comparable findings were obtained by Pratt and Whitney in their 1992 survey of 120 in-flight shutdowns occurring on Boeing 747s in 1991.[15] Here, the top three contributing factors were missing parts, incorrect parts and incorrect installation. In a UK Civil Aviation Authority survey of maintenance deficiencies of all kinds, the most frequent problem was the incorrect installation of components, followed by the fitting of wrong parts, electrical wiring discrepancies and loose objects (tools and so on) left in the aircraft.[16]

The Prevalence of Omissions

What type of error is most likely to occur during maintenance-related activities and most especially during the installation task? As noted earlier, the answer is omissions: the failure to carry out necessary parts of the task. Omissions can involve either the failure to replace some component or the failure to remove foreign objects (tools, rags and the like) before leaving the job.

Jens Rasmussen[17] analysed 200 significant event reports published in *Nuclear Power Experience* and identified the top four error types as follows.

- Omission of functionally isolated acts (34%)
- Other omissions (9%)
- Side-effect(s) not considered (8%)
- Manual variability, lack of precision (5%).

He also identified the activities most often associated with omissions, as listed below.

- Repair and modification (41%)
- Test and calibration (33%)
- Inventory control (9%)
- Manual operation and control (6%).

The INPO investigation, mentioned earlier, found that 60 per cent of all human performance root causes involved omissions and that 64.5 per cent of the errors in maintenance-related activities were omissions.[18] This study also observed that 96 per cent of deficient procedures involved omissions of one kind or another.

I analysed the reports of 122 maintenance lapses occurring within a major airline over a three-year period. A preliminary classification yielded the following proportions of error types:

- Omissions (56%)
- Incorrect installations (30%)
- Wrong parts (8%)
- Other (6%).

What gets omitted? A closer analysis of the omission errors produced the following results:

- Fastenings undone/incomplete (22%)
- Items left locked/pins not removed (13%)
- Caps loose or missing (11%)
- Items left loose or disconnected (10%)
- Items missing (10%)
- Tools/spare fastenings not removed (10%)
- Lack of lubrication (7%)
- Panels left off (3%).

It seems unnecessary to labour the point any further. Omissions represent the largest category of maintenance-related errors, and maintenance-related errors constitute the largest class of human performance problems in nuclear power plants, and probably in aviation as well—although there are no comparable data to support this view as yet.

Omission-prone Task Features

From an analytical point of view there are at least two approaches towards a better understanding of maintenance omissions, one seeking to identify the underlying cognitive mechanisms, the other trying to determine what aspects of a task cause it to be especially omission-prone. The former route is made difficult by the fact that an omission can arise within a number of cognitive processes concerned with planning and executing an action, as summarized in Table 5.3. Even when the omission is one's own, the underlying mechanisms are not easy to establish, but when the omission is made by another person at some time in the past, the underlying reasons may be impossible to discover. The task analysis route, on the other hand, is more promising.

Table 5.3   Summary of the possible cognitive processes involved in omitting necessary steps from a task

Level of failure	Nature of failure	Failure type
Planning	(a) A necessary item is unwittingly overlooked.	Mistake
	(b) The item is deliberately left out of the action plan.	Violation
Intention storage	The intention to carry out the action(s) is not recalled at the appropriate time.	Lapse
Execution	The actions do not proceed as intended and a necessary item is unwittingly omitted from the sequence.	Slip
Monitoring	The actor neither detects nor corrects the prior omission.	Slip

The act of duplicating a looseleaf document on a simple photo-copying machine gives an everyday illustration of omission-prone task steps (see Figure 5.2). There is strong evidence to show that the most likely omission is to leave the last page of the original under the lid when departing with the copy and the remainder of the original pages.

There are at least four distinct features of the photocopying task that combine to make this omission highly likely—irrespective of who is carrying out the task.

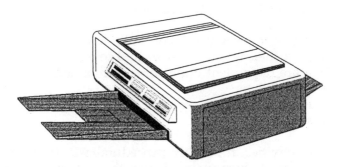

Figure 5.2 **A simple photocopier in which there is a strong likelihood of failing to remove the last page of the original**

- The step is functionally isolated from the preceding actions. Before, the act of removing the previously copied page had been cued by the need to replace it with the next page. In this instance, there is no next page.
- The need to remove the last page of the original occurs after the main goal of the activity has been achieved—obtaining a complete copy of the document—but before the task itself is complete.
- The step occurs close to the end of the task. Studies of absentminded slips in everyday life have shown that such 'premature exits' are a common form of omission which can be prompted by a preoccupation with the next task. However, in maintenance work organized around an eight- or twelve-hour shift pattern, there is no guarantee that the individual who starts upon a job will be the one to complete it. And even when the same person performs the whole task, there is always the possibility that he or she may be called away or distracted before the task is finished.
- The last page of the original is concealed under the lid of the photocopier—the out-of-sight-out-of-mind phenomenon.

To this list can be added several other features which, if present within a given task step, can combine to increase the probability that the step will be omitted. Other omission-provoking features include the following:

- Steps involving actions or items not required in other very similar tasks.
- Steps involving recently introduced changes to previous practice.

- Steps involving recursions of previous actions, depending upon local conditions.
- Steps involving the installation of multiple items (for example, fastenings, bushes, washes, spacers and so on.)
- Steps that are dependent upon some former action, condition or state.
- Steps that are not always required in the performance of this particular task.

Maintenance activities are highly proceduralized. It is therefore possible, in principle, to identify in advance those steps most vulnerable to omissions by establishing the number of omission-provoking features that each discrete step possesses. Having identified error-prone steps, remedial actions can then be taken to reduce the likelihood of these steps being left out.

The Characteristics of a Good Reminder

Although there are a variety of cognitive processes that could contribute to an omission, and their precise nature is often hidden from both the actor and the outside observer, the means of limiting their future occurrence can be relatively straightforward and easy to apply once the error-prone steps have been identified. The simplest countermeasure is an appropriate reminder. What characteristics should a good reminder possess? Some suggestions are listed below.

- It should be able to attract the actor's attention at the critical time *(conspicuous)*.
- It should be located as closely as possible in both time and distance to the to-be-remembered (TBR) task step *(contiguous)*.
- It should provide sufficient information about when and where the TBR step should be carried out *(context)*.
- It should inform the actor about what has to be done *(content)*.
- It should allow the actor to check off the number of discrete actions or items that should be included in the correct performance of the task *(check)*.

These five characteristics are universal criteria for a good reminder. They are applicable in virtually all situations. There are, however, a number of secondary criteria that could also apply in many situations:

- It should work effectively for a wide range of TBR steps *(comprehensive)*.

- It should (when warranted or possible) block further progress until a necessary prior step has been completed *(compel)*.
- It should help the actor to establish that the necessary steps have been completed. In other words, it should continue to exist and be visible for some time after the performance of the step has passed *(confirm)*.
- It should be readily removable once the time for the action and its checking have passed—one does not, for example, want to send more than one Christmas card to the same person *(conclude)*.

The presence of reminders is not a guaranteed solution to the omission problem. But—in the spirit of *kaizen*[19]—it will certainly help to bring about a substantial reduction in their numbers. Consider, for example, what the impact of the reminder shown in Figure 5.3 might be upon the likelihood of you leaving behind the last page of the original.

Figure 5.3   An example of a simple reminder to minimize the last page omission

In maintenance, any such reduction in the largest single category of human error could have substantial benefits, both in lives and in costs. According to a Boeing analysis, maintenance and inspection failures ranked second only to controlled flight into terrain in the list of factors contributing to onboard fatalities, and caused the deaths of 1481 people in 47 accidents between 1982–91.[20] However, the costs of maintenance failures are more likely to be counted in terms of money rather than casualties. Such losses can be very high. One major airline has reckoned its annual losses due to maintenance lapses at around $38 million dollars.[21] It has been estimated that an inflight engine shutdown, for example, can cost up to $500 000 in lost income and repairs; each flight cancellation can cost up to $50 000; each hour of delay on the ramp can cost $10 000.[22]

It should be noted that the reminders described above are not a permanent solution to the omission problem. They are, at best, 'first aid' measures to cope with the difficulties experienced in the present generation of hazardous systems—whose working lives will run for many years into the future. A more lasting solution would be to design components so that they can only be installed in the correct way. Another would be to make the system disable itself automatically when it detected the presence of missing parts. A third and more fundamental solution would be to design out the need for 'hands on' human contact during maintenance inspections. As a first step in the direction of the last possibility, we will review some of the engineering and economic reasons why maintenance is carried out in the first place.

The Rationale for Maintenance

Why not design systems for zero maintenance? Why not have all the constituent parts of a system lasting for exactly the same length of time—a period equal to the planned life of the system as a whole? The standard answer is that this would be uneconomical, given the complexity and expense of a modern technological system.[23] Thus, many of the components will have been designed with a useful life that is greater than the longest production cycle but less than the planned lifetime of the total system. These 'weak' components will have been identified at the design stage and made accessible and replaceable. This is termed the 'expected maintenance load'. Yet other components will fail in an unplanned way as the result of design mistakes or operational mishandling. This is the 'unexpected maintenance load'. The two are interrelated. Failure to deal with the expected maintenance load will generate a larger and more costly investment of maintenance resources to cope with unexpected failures. This interaction is summarized in Figure 5.4.

Once the 'resource elbow' is passed, the costs of restoring the unit—and hence the system—to an acceptable level accelerate dramatically. Working in the preventive maintenance zone has several advantages: it is more economical, it takes care of the expected maintenance load, and—by forestalling failures—it greatly reduces the unexpected load as well. This is the orthodox engineering view. It is summarized in Figure 5.5 that shows how the optimum level of preventive maintenance can be determined by summing the costs of both corrective and preventive maintenance, and then selecting the level coinciding with the lowest overall maintenance cost.

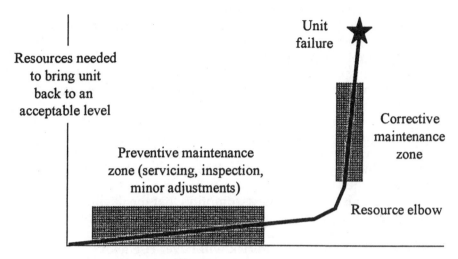

Figure 5.4   **Deterioration characteristics of a simple mechanical item (after Kelly)**

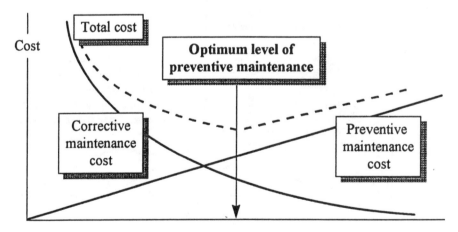

Figure 5.5   **The relationship between the level of preventive maintenance and the total maintenance cost—shown by the dotted line**

Unfortunately, this orthodoxy presumes that all—or at least most—maintenance activities are essentially benign. But suppose preventive maintenance did not always prevent failure and that corrective maintenance did not always correct it. Suppose that both of these activities actually had the potential for doing serious harm, rendering previously reliable components inoperable or simply removing them altogether.

Figure 5.6 looks at the maintenance issue from a broader perspective—one that includes human as well as technical factors. Here are plotted (in a very speculative fashion) the risks to the system posed by (a) neglected maintenance, and (b) by the likelihood of errors being committed during either preventive or corrective maintenance. The latter plot is based on the assumption that the likelihood of error will increase as a direct linear function of the amount of maintenance activity. Since only a relatively small proportion of human actions are erroneous, the human failure risk will never rise above a fairly low value. But, as we shall see below, it is not the absolute value that matters, but the relative proportions of the maintenance neglect and

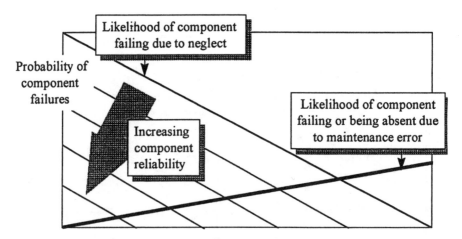

Level of maintenance activity (amount of direct human contact)

Figure 5.6   Comparison of the risks to the system of component failure due to (a) neglected maintenance and (b) errors committed during maintenance
The family of diagonal lines advancing to the lower left-hand corner reflect the increasing reliability of components over time.

maintenance error risks. It is also assumed that these error risks will not change in any systematic fashion over time. Technology may advance, but human fallibility stays the same.

In sharp contrast, however, the risks due to maintenance neglect are likely to diminish steadily as manufacturing techniques and the intrinsic reliability of materials improve with technological developments. This is indicated in Figure 5.6 by the set of dotted diagonals advancing towards the lower left-hand corner of the graph, where each dotted line represents a stage in the improvement of the technology. It is clear that if a given level of maintenance—determined by the economic and engineering considerations discussed above—remains relatively constant over time, then a point will soon be reached when the dangers to the system come to be dominated by even a relatively low error rate.

The data reported earlier on the causes of inflight engine shutdowns show that all of the most common contributing factors are associated with human rather than 'unaided' technical failures. Of course, it could be argued that the advent of non-destructive testing and other advanced diagnostic techniques allow aircraft engineers to identify potential technical failures before they happen inflight, thus leaving human errors as the main residual category of failure. This may well be true, but it does not alter the fact that regular human contact with the 4–6 million removable parts on a modern aircraft poses an unacceptable level of risk.

Ironically, one of the pressures that sustains this high level of maintenance contact is the safety-criticality of these systems. A catastrophic breakdown is unacceptable in commercial aviation, nuclear power generation or chemical process plants. Everything must be done—and be seen to be done—to preserve the integrity and reliability of these systems. But, as we have seen, the maintainer's touch can harm as well as heal, and the point seems to have been reached in some modern systems when the risks of the former may outweigh the benefits of the latter.

Conclusions

Rapid technological advances in hazardous systems have not only brought about the replacement of human control by computers, they have also led to very substantial improvements in the reliability of equipment and components. This has been achieved by the use of better manufacturing processes and materials, as well as through the widespread availability of sophisticated diagnostic techniques. But the maintenance schedule for a modern aircraft or nuclear power plant still demands the repeated disassembly, inspection and replacement of millions of removable parts over the long working life of the system. Thirty or even twenty years ago, these inspections would probably have resulted in the frequent detection and replacement of failed components. Then, the risks of failure due to intrinsic engineering defects almost certainly exceeded the dangers created by allowing legions of fallible people direct access to the vulnerable entrails of the system. But now the balance has tipped the other way.

The greatest hazard facing modern technologies comes from people, and most particularly from the well intentioned, but often unnecessary, physical contact demanded by outdated maintenance schedules. We urgently need a greater awareness on the part of system designers and manufacturers of the varieties of human fallibility and the error-provoking nature of large parts of the maintenance task—especially during installation or reassembly. Most of all, they must appreciate that maintenance can be a serious hazard as well as a necessary defence. Until systems are designed and built with these issues in mind, good maintenance personnel will go on contributing to bad organizational accidents.

Notes

1   *Encyclopaedia Britannica*, Macropaedia, vol. 17, p. 367, 1980.
2   T.A. Kletz, *What Went Wrong: Case Histories of Process Plant Disasters*, (Houston: Gulf Publishing Company, 1985).
3   J. Kemeny, *The Need for Change: The Legacy of TMI. Report of the Presidential Commission at Three Mile Island*, (New York: Pergamon, 1979).
4   M. Job, *Air Disaster*. Vol 2, (Weston Creek, ACT: Aerospace Publications Pty Ltd, ch. 4, 1996), pp. 47–60.
5   See sources cited in ch. 3, p. 46. For two contrasting accounts of the same event, see also Union Carbide, *Bhopal Methyl Isocynate Incident Investigation Team Report*, (Danbury: Union Carbide Corporation, March 1985) and W. Morehouse and M.A. Subramanian, *The Bhopal Tragedy*, (New York: Council on International and Public Affairs, 1986). See also P. Shrivastava, *Bhopal: Anatomy of a Crisis*, (Cambridge MA: Ballinger Publishing Company, 1987). For a very perceptive comparative account, see also N. Meshkati, 'An etiological investigation of micro- and macro-economic factors in the Bhopal disaster: lessons for industries of both industrialized and developing countries', *International Journal of Industrial Ergonomics*, 4, 1989, pp. 161–75, and N. Meshkati, 'Human factors in large-scale technological systems' accidents: Three Mile Island, Bhopal, Chernobyl', *Industrial Crisis Quarterly*, 5, 1991, pp. 133–54.
6   M. Job, *Air Disaster*, Vol. 2, op. cit., ch. 10, pp. 136–53.
7   The Hon. Lord. Cullen, *Public Inquiry into the Piper Alpha Disaster*, (Department of Energy. London: HMSO, 1990).
8   A. Hidden, *Investigation into the Clapham Junction Railway Accident*, (Department of Transport. London: HMSO, 1989). See also C. Perin, 'British Rail: the case of the unrocked boat', commentary given to a workshop on Managing Technological Risk in Industrial Society, 14–16 May 1992 at Bad Hamburg.

9   Occupational Safety and Health Administration (OSHA). *The Phillips 66 Company Houston Chemical Complex Explosion and Fire*, (Washington, DC: US Department of Labor (OSHA), 1990).

10   INPO, *An Analysis of Root Causes in 1983 Significant Event Reports. INPO 84-027*, (Atlanta, GA: Institute of Nuclear Power Operations, 1984).

11   INPO, *An Analysis of Root Causes in 1983 and 1984 Significant Event Reports*, INPO 85-027.

12   K. Takano, Personal communication, 1996.

13   The distinction between knowledge-in-the-world and knowledge-in-the-head comes from Don Norman's excellent book *The Psychology of Everyday Things* (New York: Basic Books, 1988).

14   Boeing, *Maintenance Error Decision Aid*, (Seattle: Boeing Commercial Airplane Group, 1994).

15   Pratt and Whitney, *Open Cowl*, March issue, 1992.

16   United Kingdom Civil Aviation Authority (UK CAA), 'Maintenance error', *Asia Pacific Air Safety*, September, 1992.

17   J. Rasmussen, 'What can be learned from human error reports?' in K. Duncan, M. Gruneberg and D. Wallis (eds), *Changes in Working Life*, (London: Wiley, 1980).

18   INPO 85-027, op cit.

19   A Japanese word meaning 'A process of deliberate, patient, continual refinements'. See Michael Crichton's novel *Rising Sun* (p. 214) for an account of the word's deeper meaning.

20   R.A. Davis, 'Human factors in the global market place. Keynote Address', *Annual Meeting of the Human Factors and Ergonomics Society*, Seattle, WA, 12 October, 1993.

21   Boeing, op. cit.

22   R.C. Graeber, 'The value of human factors awareness for airline management', Paper presented to conference on Human Factors for Aerospace Leaders, Royal Aeronautical Society, London, 28 May, 1996.

23   A. Kelly, *Maintenance Planning and Control*, (London: Butterworths, 1984).

6 Navigating the Safety Space

Assessing Safety

This chapter deals with the principles underlying the measurement of safety (the actual measures and their application will be discussed in the next chapter). In particular, it is about assessing the 'safety health' of complex technological systems that—by virtue of their many-layered defences—have relatively few bad accidents. We will also be returning to a distinction first introduced in Chapter 4—that between *process* and *outcome*.

When it comes to restricting human behaviour to safe and productive pathways, most organizations favour process controls, based upon rules, regulations and procedures (see Chapter 4). But, in the management of system safety, the reverse is the norm. Most organizations involved in hazardous operations rely heavily upon outcome measures, or, more specifically, upon negative outcome measures that record the occurrence of adverse events such as fatalities and lost time injuries.

Unfortunately, such outcome data provide an unreliable indication of a system's intrinsic safety. This is especially the case when the number of adverse events has fallen to some low asymptotic value around which the small fluctuations from one accounting period to the next are more 'noise' than 'signal'. In many well defended systems, these data are too few and too late to guide effective safety management. Indeed, some modern technologies have become the victims of their own success: they have largely eliminated the conventional outcome measures by which they were accustomed to steer system safety. In commercial aviation, for example, the accident risk has remained fairly constant over the last 25 years at an average worldwide value of one passenger fatality for every million air miles.[1]

If we are to make progress, we need to reconsider the nature of safety. Dictionary definitions are of little help. Most equate safety with freedom from danger or risk, but both are ever-present in hazardous technologies. The most widely used indicator is the number of negative outcomes, but it is only helpful when the accident rates are high enough, and even then we are left with the problem of chance—the fortuitous combination of causal elements at a particular place and time.

Only if the managers of a system had complete control over all possible accident-producing factors could negative outcome data provide a valid index of its intrinsic safety. Only then could accident rates be linked directly to the quality of safety management. But no hazardous technology can ever achieve this total control. Natural hazards can be defended against, unsafe acts can be moderated to some degree, but neither can be eliminated altogether. Latent conditions, or pathogens, will always be present. The likelihood of their adverse conjunction is always greater than zero. The large random component in accident causation means that 'safe' organizations can still have bad accidents, and 'unsafe' organizations can escape them for long periods. Bad luck can bring down the deserving, while good luck can protect the unworthy.

One way out of this impasse is through the use of *process measures* of safety. To apply such measures effectively, we need to recognize that safety has two faces, a negative one and a positive one. As with health, the occasional absences of safety are easier to quantify than its more enduring presence. In order to get closer to the positive face of safety, we need to think about horse kicks.

Counting Horse Kicks

In the early part of the nineteenth century, Siméon Poisson, the great French mathematician, counted the number of horse kicks received by Prussian cavalrymen over a given time period.[2] The exact numbers do not matter, but Figure 6.1 gives an approximate idea of what he found.

By far the largest proportion of the regiment suffered no kicks at all. A small number were kicked once, an even smaller number twice and still fewer were kicked three or more times. On the basis of these and similar data, Poisson developed a theoretical model—the Poisson distribution—for determining the chance probability of an accident among a group of people sharing equal exposure to a particular hazard. The Poisson distribution addresses the question: how many people would we expect to find with 0, 1, 2 or more, accidents when there is no special reason why one person should have more than any other?

Even in the relatively straightforward matter of horse kicks, it is very unlikely that liability to these painful encounters would have

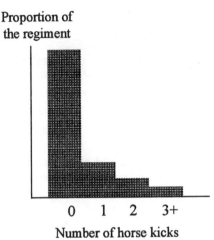

Figure 6.1   An imaginary distribution of the number of horse
             kicks suffered by a regiment of cavalry over a given
             time period

been the same for every person. It can be seen from Figure 6.1, that most of the regiment fell into the 'zero kick' category. However, there could be many different reasons for this. Some of the unkicked cavalrymen could have been just lucky. Some could have established a good relationship with their mounts. Others could have exercised greater caution. Still others could have been both kind to their horses and wary in the kicking zone. Similar, but opposite, reasons could explain why some people were kicked more often than others. We can therefore think of a continuum running from the kick-proof to the kick-prone, or from relatively safe to relatively unsafe individuals.

Many different factors act to locate an individual's position along this safe–unsafe dimension. In some cases, it will simply be a question of good or bad luck. But in others it will be due to the success of deliberate countermeasures. The most kick-resistant cavalrymen will be those who employ the best protective measures (kindness, caution and so on) in the most sustained and effective way. In short, we can discriminate degrees of safety as well as 'unsafety', as shown in Figure 6.2.

The horse kicks exercise demonstrates that outcome data can only reveal the negative face of safety. Such measures record moments of vulnerability; they cannot discriminate the more enduring ingredients of resistance. As we shall see later, these need to be assessed directly using process measures—indices that gauge the general 'safety

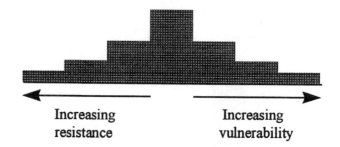

Increasing
resistance

Increasing
vulnerability

Figure 6.2   A two-sided distribution of resistance–vulnerability among kick-free cavalrymen achieved by discriminating according to the effectiveness of their countermeasures

health' of the system as a whole by sampling various vital signs on a regular basis. Only in this way can we navigate the safety space (see below) in a principled manner rather than remaining at the mercy of the prevailing currents.

Introducing the Safety Space

The safety space is a natural extension of the resistance–vulnerability continuum introduced in the previous section. It is a notional space within which we can represent the current resistance or vulnerability of an individual or an organization. As shown in Figure 6.3, it is cigar-shaped, with extreme resistance located at the left-hand end and extreme vulnerability at the right-hand end. The shape of the space acknowledges that most people or organizations will occupy some intermediate point within this space. Hereafter, we will focus upon organizations rather than individuals.

An organization's position within the safety space is determined by the quality of the processes used to combat its operational hazards. In other words, its location on the resistance–vulnerability dimension will be a function of the extent and integrity of its defences at any one point in time. But there is no such thing as absolute safety. So long as natural hazards, human fallibility, latent conditions and the possibility of chance conjunctions of these accident-producing factors continue to exist, then even the most intrinsically resistant organizations—those at the extreme left-hand end—can still have accidents. By the same token, 'lucky' but unsafe organizations at the extreme right-hand end of the space can still escape accidents for quite long periods of time.

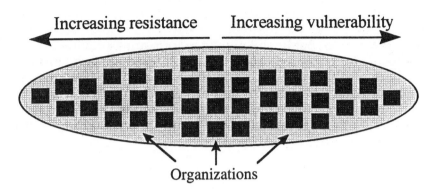

Figure 6.3 The safety space
The figure shows how 40 hypothetical organizations could be distributed within the safety space at any one time. Their positions within the space are determined by their intrinsic resistance or vulnerability to their operating hazards.

Organizations can, of course, make their own luck to some degree—but never completely. This has two implications for outcome measures. First, there will be some correspondence between an organization's position within the safety space and the number of negative outcomes that befall it. Organizations at the resistant end of the space are likely to suffer fewer bad events within a given sampling period than those at the vulnerable end. But this correlation will never be perfect due to unforeseeable interventions of happy and unhappy chance. Second, when accident rates within a particular domain fall to very low levels, as they have in aviation and nuclear power generation, the occurrence or not of negative outcomes (within a particular accounting period) reveal very little about an organization's position within the safety space—yet such differences in location will continue to exist.

Currents within the Safety Space

Very few organizations occupy fixed positions within the safety space. Most of them are in continuous motion, either being actively driven towards the resistant end of the space by the energetic implementation of effective safety measures, or by being allowed to drift passively towards the unsafe end. Naturally occurring currents are running in opposite directions within the space, their force becoming stronger the nearer an organization approaches either end. These currents are shown in Figure 6.4. The closer an organization drifts towards the unsafe end of the space, the more likely it is to suffer accidents. These, together with public and regulatory pressures, provide a powerful impetus for enhanced safety measures. Although accidents are relatively unreliable indicators of intrinsic safety, they nevertheless succeed, temporarily at least, in concentrating the minds of top management upon protection rather than production issues. This raises the question to be discussed at a later point: Do organizations need bad accidents in order to survive?

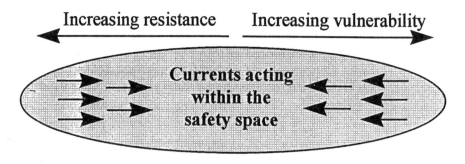

Figure 6.4   Countervailing currents within the safety space

Contrary forces will come into play as the organization moves into the more resistant regions of the space. Safety initiatives run out of steam and yield diminishing returns. Managers once again forget to be afraid and begin to divert more of their attention and resources to production goals. And, as discussed in Chapter 1, new and improved defences are used for furthering productive, rather than protective, goals. In other words, organizations become accustomed to their apparently safe state and allow themselves to drift—like the unrocked boat—into regions of greater vulnerability.

If they were to rely entirely on outcome measures, organizations would probably move passively to and fro across the central zone like pieces of flotsam. Effective safety management means actively navigating the safety space in order to reach and then remain within the zone of maximum resistance. To do this, managers must understand the nature of the forces acting upon the organization, as well as the kinds of information needed to fix their current position. To reach the target region and then stay there, two things are necessary: an internal 'engine' to drive the organization in the right direction, and 'navigational aids' to plot their progress.

What Fuels the 'Safety Engine'?

Three ingredients are vital for driving the safety engine, all of them the province of top management—or what the organizational theorist, Mintzberg, has termed the *strategic apex* of the system.[3] These driving forces are: commitment, competence and cognisance—'the three Cs'.

Commitment has two main components: motivation and resources. The motivational issue relates to whether an organization seeks to be the domain model for good safety practices or whether it is simply content to keep one step ahead of the regulators—the difference noted in Chapter 2 between generative and pathological organizations. High levels of commitment are comparatively rare and hard to sustain. This is why the organization's safety culture is so important. Top management come and go. More organizational leaders are appointed to revive sagging commercial fortunes than to improve indifferent safety records. A good safety culture, on the other hand, is something that endures beyond these palace revolutions and so provides the necessary driving force irrespective of the inclinations of the latest CEO. The second issue concerns the resources allocated to the achievement of safety goals. This is not just a matter of money. It concerns quality as well as quantity, and has to do with the calibre and status of the people assigned to direct the management of system safety. Within some organizations, safety jobs are seen as being in the fast lane of career advancement. In all too many companies, however, they are regarded as long-term parking areas for underpowered or burned-out executives.

But commitment alone is not enough. The organization must also possess the technical competence necessary to achieve its safety goals. Competence is very closely related to the quality of the organization's safety information system. Does it collect the right information? Does it disseminate it? Does it act upon it? Paired comparison studies—examining pairs of companies matched in all respects except for safety performance—have shown that the two characteristics most likely to distinguish safe organizations from less safe ones are, firstly, top-level commitment and, secondly, the possession of an adequate safety information system.[4]

Neither commitment nor competence will suffice unless the organization has a correct awareness—or cognisance—of the dangers that threaten its operations. Two features are symptomatic of organizations lacking this necessary level of cognisance. The first is the *positional paradox*—where those at the top of the organization, possessing the largest degree of decisional autonomy, blame most of their safety problems on the personal shortcomings of those at the sharp end who, for the most part, simply follow procedures and work with the equipment provided. The second symptom is the *tick-off phenomenon*. This is something that can afflict technical managers assigned to safety jobs. They treat safety measures like pieces of equipment. They put them in place, then tick them off as another job done. Most pieces of equipment do what they are supposed to do. Switch them on and they function as specified. But safety measures involve both a product and a process. Simply implementing them is not enough. They have to be watched, worried about, tuned and adjusted. Items of equipment are nearly all product and very little process. However, safety measures are more like religion—there is a great deal of praying (process), but few miracles (product).

Cognisant organizations understand the true nature of the 'safety war'. They see it for what it really is—a long guerilla struggle with no final conclusive victory. For them, a lengthy period without a bad accident does not signal the coming of peace. They see it, correctly, as a period of heightened danger and so reform and strengthen their defences accordingly. In any case, since entropy wins in the end, a more appropriate metaphor for the conclusion of the safety war might be likened to the last helicopter out of Saigon rather than a decisive Yorktown, Waterloo or Appomatox.

Setting the Right Safety Goals

The key to navigating the safety space lies in appreciating what is manageable and what is not. Many organizations treat safety management as a negative production process. They set reduced negative outcome targets for the coming accounting period (e.g., 'Next year, we'll reduce our lost-time accidents by half'.). But accidents, by their nature, are not directly controllable. So much of their causal variance lies outside the organization's sphere of influence. The organization can only defend against hazards, it cannot remove or avoid them—and still stay in business. Similarly, an organization can only strive to minimize unsafe acts, it cannot eliminate them altogether.

Effective safety management is more like a long-term fitness programme than negative production. Rather than struggling vainly to exercise direct control over incidents and accidents, managers should regularly measure and improve those processes—design, hardware, training, procedures, maintenance, planning, budgeting, communication, goal conflicts, and the like—that are known to be implicated in the occurrence of organizational accidents. These are the manageable processes determining a system's safety health. They are, in any case, the processes that managers are hired to manage. In this way, safety management is not an add-on, but an essential part of the system's core business.

As indicated earlier, the only attainable goal for safety management is not zero accidents, but to reach that region of the safety space associated with maximum resistance—and then staying there. Simply moving in the direction of greater safety is not difficult. But sustaining these improvements is very hard. To hold such a position against the strong countervailing currents requires navigational aids. More specifically, it requires a safety information system that is not only capable of revealing the right conclusions about past events (reactive measures), but that also facilitates regular 'health checks' of the basic organizational processes (proactive measures) which are then used to guide targeted remedial actions. The navigational aids are shown in Figure 6.5 and discussed further in succeeding sections.

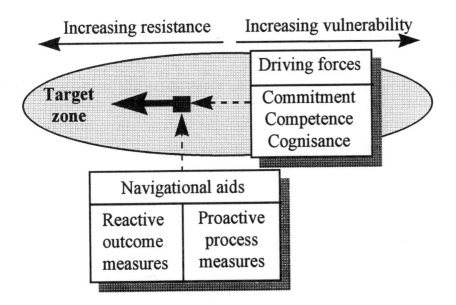

Figure 6.5 **A summary of the principal factors involved in navigating the safety space:**
(a) the driving forces, and (b) the navigational aids that together comprise the safety information system.

A Test to Destruction

In case the ideas of resistance or vulnerability to hazards seem too abstract, consider the results of the following study. One way of determining the relative resilience of different systems is to establish how many things went wrong before they sustained a serious accident. In effect, this is the equivalent of an engineering test-to-destruction.

The systems in this case were types of aircraft: helicopters, light aircraft (general aviation) and large commercial jets. The data source was 90 fatal accident investigations carried out by the UK Air Accident Investigation Branch between the 1970s and the 1990s.[5] All such investigations were conducted and reported in the standardized fashion laid down by the International Civil Aviation Organization's guide-lines for accident investigators.

Sixteen possible contributing factors were identified in advance: engine problems, airframe problems, system problems, fuel problems, pilot hours on aircraft type, rest periods, pilot error, operator issues, change of plan, time of day/night, icing conditions, visibility, wind, precipitation, air traffic control and radio communications. The number of such problems was established for each fatal accident, and the results are summarised in Table 6.1. Not surprisingly, helicopters were the most vulnerable aircraft type. In contrast, it required more than twice the number of contributing factors to bring about a fatal accident in the better defended commercial jet airliners.

Table 6.1   **Mean numbers of problems contributing to fatal accidents in three aircraft types**

Aircraft type	Number of accidents	Mean number of problems
Helicopters	22	1.95
Light aircraft	29	3.38
Large jets	39	4.46

Source: Data from UK AAIB reports.

An Overview of the Navigational Aids

Navigational aids fall into two main categories: *reactive measures* that can only be applied after the occurrence of an event, and *proactive measures* that can be used before an event to assess the 'safety health' of the system as a whole. Effective safety management requires the use of both of these measures. We need to learn the right lessons from past events, and then translate this knowledge into enhanced resistance. At the same time, we must make visible to those who manage and operate the system the latent conditions and resident pathogens that are an inevitable part of any complex technology, no matter how well managed, and that experience has shown can provide the ingredients for future organizational accidents.

Used properly, both reactive and proactive measures can give valuable information about the state of the underlying organizational processes. The accident causation model presented in Figure 1.6 directs our attention to two important issues: the local and organizational conditions that promote unsafe acts, and the barriers, safeguards and defences that keep hazards and potential losses apart. Table 6.2 summarizes their various interactions.

Table 6.2 **Summary of the possible interactions between reactive and proactive measures**

	Type of navigational aid	
	Reactive measures	**Proactive measures**
Local and organizational conditions	Analysis of many incidents can reveal recurrent patterns of cause and effect.	Identify those conditions most needing correction, leading to steady gains in resistance or 'fitness'.
Defences, barriers and safeguards	Each event shows a partial or complete trajectory through the defences.	Regular checks reveal where holes exist now and where they are most likely to appear next.

One of the commonest misuses of reactive measures is to focus too narrowly upon single events. This leads to countermeasures aimed chiefly at preventing the recurrence of individual failures, particularly human ones. But organizational accidents have multiple causes. No one factor is necessarily more important than any other. It was their combination that caused the event.

To learn the right lessons from the past, it is best to analyse several domain-related events using a common classificatory framework. This reveals patterns of cause and effect that are rarely evident in single-case investigations. Moreover, these patterns also indicate which of the local or organizational factors is playing a regular part in contributing to adverse events. This, in turn, guides the selection of which organizational processes to sample proactively on a regular basis. There are many possible candidates for regular assessment, but few organizations have the resources to sample them all, nor would it be appropriate to do so. The analysis of multiple events tells us which of many possible processes are the most likely to provide a valid and cost-effective measure of current safety health.

An effective reactive technique should give an accurate picture of the weaknesses and absences that existed in the defences. These 'snapshots' reveal the often improbable ways that even the most elaborate defences can be defeated. Once again, analyses of several events from the same domain can bring to light recurrent, and often surprising, patterns of defensive weakness.

The most important navigational aids, however, lie on the right-hand side of Table 6.1. They are the proactive process measures that can be applied before a bad event. Their purpose is to provide regular checks both on the system's defences and on its 'vital signs' at the workplace and organizational levels. Just as in medicine, there is no single comprehensive measure of safety health. It involves sampling a subset of a potentially larger collection of indices reflecting the current state of various organizational processes. The number of such diagnostic checks ranges typically from around eight to 16 and will vary from one type of system to another and their purpose is to identify those two or three processes that are in most urgent need of attention. The frequency with which these checks are carried out will depend upon the rate at which the process in question is likely to change. Workplace factors change more rapidly than organizational ones, and so need to be monitored at more frequent intervals.

We will return to these procedures at a later point, and a practical guide to their selection and application will be given in the next Chapter. In the meantime, we must look briefly at near-miss and incident reporting, a type of safety measurement that is not only developing rapidly in a variety of domains, but which also yields both reactive and proactive data.

Near-miss and Incident Reporting Schemes

For some people, the terms 'near-miss' and 'incident' have distinct meanings, but since this is not a universal practice, we will use 'near-miss' to cover all such events. A near-miss is any event that could have had bad consequences, but did not. Near-misses can range from a partial penetration of the defences to situations in which all the available safeguards were defeated but no actual losses were sustained. In other words, they span the gamut from benign events in which one or more of the defences prevented a potentially bad outcome as planned, to ones that missed being catastrophic by only a hair's breadth. The former provide useful proactive information about system resilience, while the latter are indistinguishable from fully-fledged accidents in all but outcome, and so fall squarely into the reactive camp.

The advantages of collecting and analysing near-misses are clear. In Exxon's nice phrase, they provide 'free lessons'.[6]

- If the right conclusions are drawn and acted upon, they can work like 'vaccines' to mobilize the system's defences against some more serious occurrence in the future—and, like good vaccines, they do this without damaging anyone or anything in the process.
- They provide qualitative insights into how small defensive failures can line up to create large disasters.
- Because they occur more frequently than bad outcomes, they yield the numbers required for more penetrating quantitative analyses.
- And, perhaps most importantly, they provide a powerful reminder of the hazards confronting the system and so slow down the process of forgetting to be afraid. But, for this to occur, the data need to be disseminated widely, particularly among the bean counters in the upper echelons of the organization. The latter have been known to become especially alert when the information relating to each event includes a realistic estimate of its potential financial cost to the organization.

There are, of course, a number of problems facing near-miss reporting schemes, not least that they all depend upon the willingness of individuals to report events in which they themselves may have played a significant part. The factors contributing to a 'reporting culture' are discussed at length in Chapter 9.

An informant may be willing to report an event, but not be able to give a sufficiently detailed account of the contributing factors. Sometimes reporters are not aware of the upstream precursors. On other occasions, they may not appreciate the significance of the local workplace factors. If they have been accustomed to working with substandard equipment, they may not report this as a contributing factor. If they habitually perform a task that should have been supervised but was not, they may not recognize the lack of supervision as a problem—and so on.

Together, the willingness and the ability issues are likely to have two effects: not all near-misses will be reported, and the quality of information for any one event may be insufficient to identify the critical precursors. But the very considerable successes of the best schemes (discussed in Chapter 7) indicate that the advantages greatly outweigh these difficulties. There seems little doubt that near-miss reporting schemes can provide an invaluable navigational aid.

Proactive Process Measurement: The Priorities

The rationale for making regular assessments of the organizational processes underlying both safety and quality has already been given above. Our purpose here is to look more closely at the things that could be measured and the likely returns on investment for each of the possible areas. The areas in question are those outlined in the triangular region at the bottom of Figure 1.6, and shown separately in Figure 6.6. They are unsafe acts, local workplace factors and organizational factors. All three exist independently of events. Defences are not considered as a separate entity because they are so closely intermingled with the human, local and organizational factors already identified in Figure 6.6.

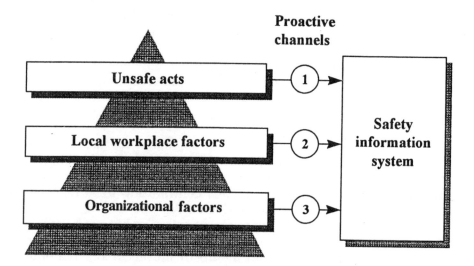

Figure 6.6   The candidate areas for proactive process measurement, each with a separately numbered channel to the safety information system

The relative merits of each type of proactive process measurement are considered below:

- *Unsafe acts.* Unsafe acts are the stuff of which accidents are made. The numerous 'iceberg' theories of accident causation presume various—widely differing—ratios between fatalities, injuries and unsafe acts. But there are serious measurement problems. The actual numbers of unsafe acts committed are almost impossible to determine. What is certain is that these numbers are large. Of greater value than the raw numbers— more or less unobtainable in any case—is information relating to the nature and variety of unsafe acts, particularly when it is linked to the type of activity being performed (see Chapter 5). Different unsafe acts require different kinds of management (see Chapter 7). Errors are essentially information-processing problems and require the provision of better information, either in the person's head or in the workplace. Violations, on the other hand, have their origins in motivational, attitudinal, group and cultural factors, and need to be tackled by counter-measures aimed more at the heart rather than the head.

- *Local workplace factors.* One stage upstream from unsafe acts are their immediate mental and physical precursors—such things as poor workplace design, clumsy automation, inadequate tools and equipment, unworkable procedures, the absence of effective supervision, high workload, time pressure, inadequate training and experience, unsociable hours, unsuitable shift patterns, poor job planning, undermanning, badly calibrated hazard perception, inadequate personal protective equipment, poor teamwork, leadership shortcomings and the like. They are likely to be fewer in number than the unsafe acts they breed. As such, they are more easily managed than the human condition. But they are still only the local expressions of higher-level organizational problems.

- *Organizational factors.* Only in the upper levels of the system can we begin to get to grips with the 'parent' failure types—the processes that create the downstream 'problem children'. If these remain unchanged, then efforts to improve things at the workplace and worker level will be largely in vain. The damaging effects of certain kinds of unsafe act may be reduced and specific conditions within the workplace improved, but the continuing existence of the 'parent' failures in the upper echelons of the organization will ensure their rapid replacement by other kinds of human and workplace problems. Clearly, then, organizational factors represent the priority area for process measurement. But how do we select the processes to be measured?

Organizations are made up of many elements. If each element were wholly independent of the others, it would only be possible to assess a company's overall safety health by measuring all the elements individually. Alternatively, if all the elements were closely related one to another, then the state of any one of them should provide a global indication of the organization's intrinsic safety. The reality probably lies somewhere between these two extremes, with the individual elements being clustered in an overlapping and modular fashion. A recent review of a number of safety process measures identified five broad clusters, as listed below:[7]

- *safety-specific factors* (for example, incident and accident reporting, safety policy, emergency resources and procedures, off-the-job safety and so on)
- *management factors* (for example, management of change, leadership and administration, communication, hiring and placement, purchasing controls, incompatibilities between production and protection and so on)
- *technical factors* (for example, maintenance management, levels of automation, human-system interfaces, engineering controls, design, hardware and so on)
- *procedural factors* (for example, standards, rules, administrative controls, operating procedures and so on)
- *training* (for example, formal versus informal methods, presence of a training department, skills and competencies required to perform tasks and so on).

At the core of these clusters and pervading all of them is the issue of culture. For the present, we can link cultural factors to the three Cs, discussed earlier: commitment, competence and cognisance—but as they exist within the organization as a whole, rather than in the mind of any one senior manager. This composite picture of the main dimensions of process measurement is summarized in Figure 6.7.

The message of Figure 6.7 is straightforward. There will be wide variation in the kinds of process that could be measured proactively from one domain to the next. This is not a problem so long as sufficient and even-handed attention is given to the main subsystems underpinning organizational safety. There are many possible ways in which one could assess the quality of the training, the procedures, the engineered safety features, and so on. What matters is that a principled attempt is made to sample each of the six main dimensions identified above: culture, training, management, safety-related issues, procedures and technical factors. The purpose of each measurement exercise is twofold. First, to identify those processes giving the greatest cause for concern at that time. Second, to track the effectiveness of previous remedial measures. Measurements that are not used to guide the enhancement of system 'fitness' are not worth taking.

Figure 6.7 **The primary process subsystems underlying organizational safety**
It will be seen that training is represented as a universal feature rather than as a localised cluster of related items.

Are Accidents Really Necessary?

History shows that the cause of safety flourishes in the aftermath of disaster, albeit briefly. Must we conclude that it takes catastrophes of the magnitude of *Piper Alpha*, Chernobyl, or Bhopal before politicians and top managers understand that investment in safety is good business? Do organizations have to fall over the edge before they know where it is? Are accidents, even the more frequent small ones, necessary to calibrate the effectiveness of safety measures? Are companies doomed to fighting the last fire or trying to prevent the last crash? The answer must be yes—if complex hazardous organizations continue to rely principally on outcome measures in order to navigate the safety space.

This chapter has outlined a workable alternative—the regular assessment of the organizational processes that are common to both quality and safety. Latent accident-producing conditions are present now. It is not necessary to wait for bad events to find out what they are. But we cannot expect to remedy them all at once. Systems need principled ways of identifying their most urgent process problems in order to deploy their limited remedial resources in the most efficient and timely manner.

We have discussed some of the principles by which cost-effective safety measurement can be achieved. Making and acting upon proactive assessments of the system's vital signs together with the intelligent application of near-miss reporting will not guarantee freedom from accidents, but it will take an organization closer to the only achievable safety goal—acquiring the maximum degree of intrinsic resistance to its local hazards and then sustaining it. The last being the hardest part.

In the next chapter, we move from guiding principles to real-world practices. Human errors pose the greatest single threat to hazardous technologies. Chapter 7 offers a practical guide to error management and reviews some of the measuring instruments and tools currently available for this purpose.

Notes

1   R.W. Howard, 'Breaking through the 10^6 barrier', *Proceedings of the International Federation of Airworthiness Conference*, Auckland, New Zealand, 20–23 October 1991.
2   W. Haddon, E.A. Suchman and D. Klein, *Accident Research: Methods and Approaches*, (New York: Harper & Row, 1964).
3   H. Mintzberg, *Mintzberg on Management: Inside Our Strange World of Organizations*, (New York: The Free Press, 1989).
4   See M.J. Smith, H. Cohen, A. Cohen and R.J. Cleveland, 'Characteristics of successful safety programs', *Journal of Safety Research*, **10**, 1988, pp. 5–14. See also U. Kjellen, 'An evaluation of safety information systems of six medium-sized and large firms', *Journal of Occupational Accidents*, 3, 1983, pp. 273–88.
5   D. Stephens, *The 'Rule of Three' as Applied to Aircraft Accidents*, (Manchester: Department of Psychology, University of Manchester, 1996).
6   I heard this from John Sleigh, formerly Head of the Safety Services Organization at the Esso Refinery, Fawley.
7   J. Reason, 'The dimensions of safety' in J. Patrick (ed.) *Cognitive Science Approaches to Cognitive Control*, Third European Conference, (Cardiff: University of Wales, 1991).

7 A Practical Guide to Error Management

What is Error Management?

Error management (EM) has two components: *error reduction* and *error containment*. Error reduction comprises measures designed to limit the occurrence of errors. Since this will never be wholly successful, we also need error containment—measures designed to limit the adverse consequences of those errors that still occur. At this general level, EM is indistinguishable from quality management or, indeed, from good management of any kind.

Error management includes:

- Measures to minimize the error liability of the individual or team.
- Measures to reduce the error vulnerability of particular tasks or task elements.
- Measures to discover, assess and then eliminate error-producing (and violation-producing) factors within the workplace.
- Measures to diagnose organizational factors that create error-producing factors within the individual, the team, the task or the workplace.
- Measures to enhance error detection.
- Measures to increase the error tolerance of the workplace or system.
- Measures to make latent conditions more visible to those who operate and manage the system.
- Measures to improve the organization's intrinsic resistance to human fallibility.

Ancient but often Misguided Practices

There is nothing new in the idea of error management. The rope's end and the whip were probably among its first instruments. But the implementation of a principled and comprehensive programme is very rare indeed. Most attempts at error management are piecemeal rather than planned, reactive rather than proactive, event-driven rather than principle-driven. They also largely ignore the substantial developments that have occurred in the behavioural sciences over the last 20–30 years in understanding the nature, varieties and causes of human error.[1]

Some of the problems associated with existing forms of EM include the following:

- They 'firefight' the last error rather than anticipating and preventing the next one.
- They focus on active failures rather than latent conditions
- They focus on the personal, rather than the situational contributions to error.
- They rely heavily on exhortations and disciplinary sanctions.
- They employ blame-laden and essentially meaningless terms such as 'carelessness', 'bad attitude', 'irresponsibility'—even in Total Quality Management (TQM).[2]
- They do not distinguish adequately between random and systematic error-causing factors.
- They are generally not informed by current human factors knowledge regarding error and accident causation.

Errors are Consequences not Causes

At the time of writing, I have before me a newspaper headline that reads 'Human error is blamed for crash'. This is a source of confusion rather than clarification. In aviation and elsewhere, human error is one of a long-established list of 'causes' used by the press and accident investigators. But human error is a consequence not a cause. Errors, as we have seen in earlier chapters, are shaped and provoked by upstream workplace and organizational factors. Identifying an error is merely the beginning of the search for causes, not the end. The error, just as much as the disaster that may follow it, is something that requires an explanation. Only by understanding the context that provoked the error can we hope to limit its recurrence.

So why are people so ready to accept human error as an explanation rather than as something that needs explaining? The answer is deeply rooted in human nature. Psychologists call it the *fundamental attribution error*.[3]

When we see or hear of someone performing badly, we attribute this to some enduring aspect of the individual's personality. We say that he or she is careless, silly, stupid, incompetent, reckless or thoughtless. But if you were to ask the person in question why they are behaving in this fashion, they would almost certainly point to the local situation and say they had no choice—circumstances forced them to do it that way. The reality, of course, lies somewhere in between.

The Blame Cycle

Why are we so ready to blame people rather than situations? Part of the answer lies in the *illusion of free will*.[4] It is this that makes the fundamental attribution error so basic to human nature.

People, especially in Western cultures, place great value in the belief that they are free agents, the captains of their own fate. They can even become physically or mentally ill when deprived of this sense of personal autonomy. Placing so much value, as we do, on this sense of individual freedom, we naturally assume that other people are similarly the controllers of their own destinies. They are also seen as free agents, able to choose between right and wrong, and between correct and error-prone paths of action. When people are given accident reports to read and asked to judge which causal factors were the most avoidable, they almost invariably identify the human actions. They are seen as far less constrained or fixed than any of the situational or organizational contributions. It is this, together with the illusion of free will, that drives the blame cycle shown in Figure 7.1.

Because people are regarded as free agents, their errors are seen as being, at least in part, voluntary actions. Although deliberate wrongdoing attracts warnings, sanctions, threats and exhortations not to do it again, these have little or no effect upon the error-producing factors, and so errors continue to be involved in incidents and accidents. Now the bosses are doubly aggrieved. People have been warned and punished, but they persist in making errors. Now they seem to be deliberately flouting the management's authority, and those who commit subsequent errors are given even stronger warnings and suffer heavier sanctions. And so the cycle goes on.

Of course, people can behave carelessly and stupidly. We all do at some time or another. But a stupid or careless act does not necessarily make a stupid or careless person. Everyone is capable of a wide range of actions—sometimes inspired, sometimes foolish—but mostly somewhere in between. One of the basic principles of error management is that the best people can sometimes make the worst errors. So how do we break free of the blame cycle? We must recognize the following basic facts about human nature and error.

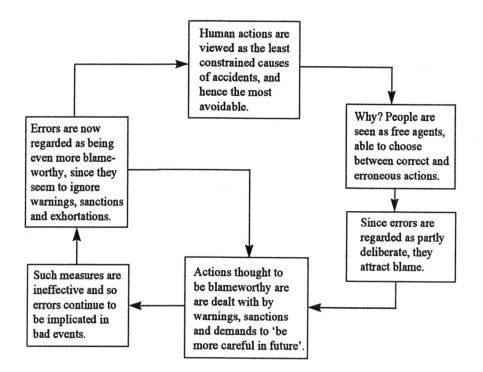

Figure 7.1   The elements of the blame cycle

- Human actions are almost always constrained by factors beyond an individual's immediate control.
- People cannot easily avoid those actions that they did not intend to perform in the first place.
- Errors have multiple causes: personal, task-related, situational and organizational factors.
- Within a skilled, experienced and largely well intentioned workforce, situations are more amenable to improvement than people.

People or Situations?

Human behaviour is governed by the interplay between psychological and situational factors. Free will is an illusion because our range of actions is always limited by the local circumstances. This applies to errors as to all other human actions. Such claims raise a crucial question for all those in the business of minimizing potentially dangerous errors: which is the easiest to remedy, the person or the situation?

Common sense and general practice would suggest that it is the person. After all, people can be retrained, disciplined, advised or warned in ways that will make them behave more appropriately in the future—or so it is widely believed. And this view is especially prevalent in professions that take pride in their willing acceptance of personal responsibility—doctors, pilots, engineers and the like. Situations, in contrast, appear as givens: we seem to be stuck with them. But is this actually the case?

Not really. The balance of scientific opinion clearly favours the situational rather than the personal approach to error management.[5] There are many reasons for this.

- Human fallibility can be moderated up to a point, but it can never be eliminated entirely. It is a fixed part of the human condition, partly because errors, in many contexts, serve a useful function (for example, trial-and-error learning in knowledge-based situations).
- Different error types have different psychological mechanisms, occur in different parts of the organization and require different methods of management.
- Safety-critical errors happen at all levels of the system, not just at the sharp end.
- Measures that involve sanctions, threats, fear, appeals and the like have only a very limited effectiveness. And, in many cases, they can do more harm—to morale, self-respect and a sense of justice—than good.
- Errors are a product of a chain of causes in which the precipitating psychological factors—momentary inattention, misjudgement, forgetfulness, preoccupation—are often the last and least manageable links in the chain.
- The evidence from a large number of accident inquiries indicates that bad events are more often the result of error-prone situations and error-prone activities than they are of error-prone people. Such people do, of course, exist, but they seldom remain at the hazardous sharp end for very long. Quite often, they get promoted to management.

An Overview of the Error Management Tool Box

It is not within the scope of this book to make a definitive review of all the available techniques that relate either directly or indirectly to error management—in the broadest sense of the term. We will not, for example, deal with the many measures that have evolved over the years and are now widely used. These include: selection, training, licensing and certification, skill checks, human resource management, quality monitoring and auditing, technical safety audits, unsafe act auditing, hazard management systems, procedures, checklists, rules and regulations, administrative controls (for example, permit-to-work systems), TQM and the like. Nor will we consider those instruments that are largely the province of reliability engineers and technical specialists. These would include probabilistic safety assessment (PSA), other event and fault tree analyses, human reliability analysis (HRA), human error analysis (HEA), hazard and operability studies (HAZOPs), failure modes and effects analysis (FMEA), and similar measures. It is not that these techniques are regarded as unimportant—on the contrary, they form an essential part of the safety manager's tool box—but they have already been discussed at length in a number of excellent recent books.[6]

Our focus here will be upon those techniques of EM that relate directly to the issues discussed throughout this book, and most particularly upon those measures that follow the principles of safety assessment set out in Chapter 6. It is hoped that what the resulting selection lacks in scope, it will make up for in novelty and practical utility. Comprehensive error management can, and should, be directed at several different levels of the organization—the individual and the team, the task, the workplace and the organizational processes. Many organizations already target most of their EM resources at the individual, as indicated earlier. For this reason, and because it is more in keeping with the book's emphasis on contextual rather than personal factors, we will not discuss individual measures any further. Nor will we deal with team-related measures—they have been covered very widely elsewhere. Over the last 15 years, the major airlines have been training their flight crews—and more recently their maintenance engineers—in Crew Resource Management (or Cockpit Resource Management). This technique has proved to be very successful in improving flight deck performance, particularly with regard to better sharing of situational awareness, improved communications and enhanced leadership skills. However, CRM techniques, their strengths and pitfalls, have already been the subject of a number of recent books, and so will not be pursued further here.[7]

That leaves only three levels for consideration: the task, the workplace and the organization. A technique for identifying error-prone steps in maintenance tasks was described in Chapter 5, and will not be repeated here. So that brings us to the main areas for consideration—reactive and proactive tools designed to reveal and correct error-producing factors at both the workplace and the organ-

Table 7.1   Summary of error management tools

Error management tools	Original domains of application	Organizational levels addressed	Reactive– proactive
Tripod-Delta	Oil exploration and production. Shipping.	Workplace and organizational factors	Proactive
Review	Railway operations	Workplace and organizational factors	Proactive
Managing Engineering Safety Health (MESH)	Aircraft maintenance	Workplace and organizational factors	Proactive
Human Error Assessment & Reduction Technique (HEART)	Potentially applicable to any hazardous operations	Task, workplace and organizational factors	Proactive
Influence Diagram Methodology (IDM)	Potentially applicable to any hazardous operations	Task, workplace and organizational factors	Reactive and proactive
Maintenance Error Decision Aid (MEDA)	Aircraft maintenance	Task, workplace and organizational factors	Reactive
Tripod-Beta	Oil exploration and production	Task, workplace and organizational factors	Reactive

izational levels. The measures to be discussed are summarized in Table 7.1.

In the descriptions that follow, Tripod-Delta will be discussed in more detail than the other proactive process-measuring instruments. There are a number of reasons for this:

- Tripod-Delta is the 'ancestor' of both Review and MESH. As such, it embodies all the principles that underlie these later techniques.
- Tripod-Delta has now been in use for several years, so we know more about its strengths and weaknesses than is the case for the other techniques.

- Tripod-Delta has been tested in a wide variety of continents, cultures, and operations, from North America and Northern Europe, through the Middle East, Africa and South-East Asia to Australia. It has also been applied successfully in a maritime setting.

Tripod-Delta

Tripod-Delta was created for the oil exploration and production operations of Shell Internationale Petroleum Maatschappij (now Shell International Exploration and Production BV) by a research team from the Universities of Leiden and Manchester.[8] The Tripod project began in 1988. The technique was developed in various Shell operating companies from 1989–1992. The first version was issued Shell-wide in 1993, and the present revised version—now known as Tripod-Delta (to distinguish it from its close relative, Tripod-Beta)—was released in 1996.

Tripod-Delta has three elements:

- A coherent safety philosophy that leads to the setting of attainable safety goals.
- An integrated way of thinking about the processes that disrupt safe operations.
- A set of instruments for measuring these disruptive processes—termed General Failure Types (GFTs)—that does not depend upon incident or accident statistics (that is, outcome measures).

During Tripod-Delta's development, Shell's principal safety metric was the LTIF, the number of lost-time injuries per million manhours. The reduction of lost-time injuries (LTIs) was the main focus for the Tripod programme. But Tripod-Delta did not seek to reduce LTIs directly. It operated at one level removed by addressing the General Failure Types, the situational and organizational factors that provoked LTIs. The underlying philosophy can be summarized as follows:

- Safety management is essentially an organizational control problem.
- The trick is to know what is controllable and what is not.
- Unsafe acts, LTIs and accidents are born from the union of two sets of parents: General Failure Types (see below for a full description) and local triggering factors.
- But only one of these parents is knowable in advance and thus potentially correctable before it causes harm: namely, GFTs or

the latent conditions associated with specific organizational processes.

- LTIs are like mosquitoes. It is pointless trying to deal with them one by one. Others simply appear in their place. The only long-term solution is to drain the swamps in which they breed—that is, the General Failure Types.
- Effective safety management depends upon the regular measurement and selective remediation of the GFTs. The Tripod-Delta instruments are designed to guide this process.

Tripod takes its name from the three-part structure illustrated in Figure 7.2. This figure also summarizes the main elements of the Tripod philosophy. The bottom right-hand 'foot' of Figure 7.2 represents the traditional concern of safety management: the performance of unsafe acts in hazardous circumstances. On occasions, these penetrate the defences to produce bad outcomes. In the past, most

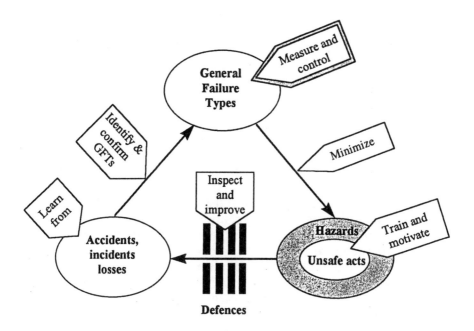

Figure 7.2   **The three 'feet' of Tripod-Delta: general failure types, unsafe acts, negative outcomes**
Also identified are the safety management actions necessary at each stage. The most important of these is to measure and control the GFTs.

remedial measures had been directed at these lower two 'feet'. Strenu-
ous efforts were made to train and motivate people to work safely.
The defences were regularly monitored and improved. And events
were investigated in order to prevent their recurrence. Tripod-Delta
adds a third and most important dimension—the measurement and
control of the GFTs. The GFTs are identified in part from the recurrent latent conditions associated with past events (indicated by the
arrow linking events to GFTs). The GFTs, in turn, create the conditions that promote or exacerbate unsafe acts (indicated by the arrow
linking GFTs with unsafe acts). The nature, measurement and control
of the GFTs are discussed below.

After observing operations in a number of operating companies and
studying their accident records, 11 GFTs were chosen as best reflecting
those workplace and organizational factors most likely to contribute to
unsafe acts and hence create LTIs. They are listed below:

- *Hardware.* This relates to the quality and availability of tools and
  equipment. Its principal components would include policies and
  responsibilities for purchase, quality of stock system, quality of
  supply, theft and loss of equipment, short-term renting, compliance to specifications, age of equipment, non-standard use of
  equipment and so on.
- *Design.* Design becomes a GFT when it leads directly to the
  commission of errors and violations. There are three main classes
  of problem: a failure on the part of the designer to provide
  external guidance (the knowledge gulf); designed objects are
  often opaque with regard to their inner workings, or to the
  range of safe actions (the execution gulf); and the failure of
  designed items to provide feedback to the user (the evaluation
  gulf).
- *Maintenance management.* This GFT is concerned with the management rather than the execution of maintenance activities
  (that are covered by other GFTs). Was the work planned safely?
  Did maintenance work or an associated stoppage cause a hazard? Was maintenance carried out in a timely fashion?
- *Procedures.* This relates to the quality, accuracy, relevance, availability and workability of procedures.
- *Error-enforcing conditions.* These are conditions relating either to
  the workplace or to the individual that can lead to unsafe acts.
  They break down into two broad (and, to a degree, overlapping)
  categories: error-producing conditions and violation-promoting
  conditions. Error-enforcing conditions, receive influences from
  many of the 'upstream' GFTs, as shown in Figure 7.3.
- *Housekeeping.* This constitutes a GFT when problems have been
  present for a long time and when various levels of the organ-

ization have been aware of them but nothing has been done to correct them. Its 'upstream' influences include: inadequate investment, insufficient personnel, poor incentives, poor definition of responsibility, poor hardware.

- *Incompatible goals.* Goal conflicts can occur at any of three levels:
  - individual goal conflicts caused by preoccupation or domestic concerns
  - group goal conflicts, when the informal norms of a work group are incompatible with the safety goals of the organization
  - conflicts at the organizational level in which there is incompatibility between safety and productivity goals.
- *Communications.* Communication problems fall into three categories:
  - system failures in which the necessary channels of communication do not exist, or are not functioning, or are not regularly used
  - message failures in which the channels exist but the necessary information is not transmitted
  - reception failures in which the channels exist, the right message is sent, but it is either misinterpreted by the recipient or arrives too late.
- *Organization.* This concerns organizational deficiencies that blur safety responsibilities and allow warning signs to be overlooked. The three main components are: organizational structure, organizational responsibilities and the management of contractor safety.
- *Training.* Problems include the failure to understand training requirements, the downgrading of training relative to operations, the obstruction of training, insufficient assessment of results, poor mixes of experienced and inexperienced personnel, poor task analyses, inadequate definition of competence requirements and so on.
- *Defences.* These comprise failures in detection, warning, personnel protection, recovery, containment, escape and rescue.

The Tripod-Delta assessments for any particular type of operation (for example, drilling, seismic, engineering, road transport, shipping and the like) are derived from checklists based upon specific indicators—or symptoms—of the presence and degree of each GFT. These indicators are obtained directly from task specialists; that is, from those involved in the day-to-day management and operation of each particular activity. Those on the spot have vital (and usually untapped) knowledge about what is safe and what is dangerous in a particular type of work. Tripod-Delta is tailored specifically for their

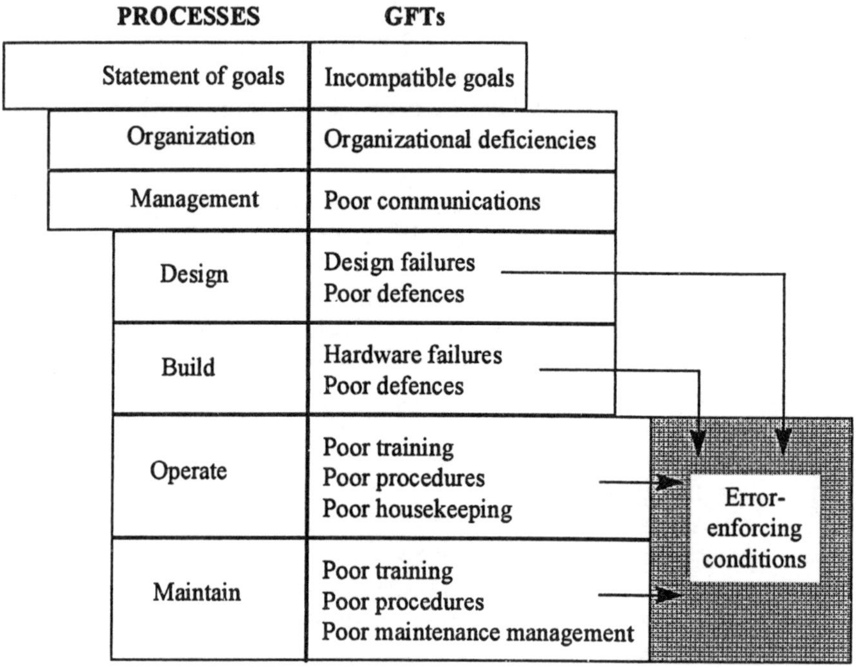

PROCESSES	GFTs
Statement of goals	Incompatible goals
Organization	Organizational deficiencies
Management	Poor communications
Design	Design failures Poor defences
Build	Hardware failures Poor defences
Operate	Poor training Poor procedures Poor housekeeping
Maintain	Poor training Poor procedures Poor maintenance management

Error-enforcing conditions

Figure 7.3   The relationships between the basic systemic processes and the general failure types, and the combined impact of the GFTs on the error-enforcing conditions

use, thus ensuring that they see the information provided as relevant to their job. This means that the instruments are built and owned by those actually carrying out the core business.

Each indicator relates to a tangible item—something that can be observed directly within the facility or can be found within a filing system—and requires a simple 'yes/no' response. Below are listed some indicator items relating to design on an offshore platform:

- Was this platform originally designed to be unmanned?
- Are shutoff valves fitted at a height of more than two metres?
- Is standard (company) coding used for the pipes?
- Are there locations on this platform where the deck and the walkways differ in height?
- Have there been more than two unscheduled maintenance jobs over the past week?
- Are there any bad smells from the low-pressure vent system?

The indicators are created by task specialists working in syndicates. The Tripod-Delta software—implemented on a PC—stores the indicator databases (one database for each GFT for each type of operation). It constructs the measuring instrument for each testing occasion by selecting 20 indicators for each GFT and then produces a 220-item checklist to be completed by a member of the workforce (for example, a drilling supervisor). Only a small proportion of items recur from one checklist to the next. The checklist can be presented either on the computer screen or in hard copy. Once completed, the data are analysed by the software which generates a Failure State Profile—a bar chart—showing the relative cause for concern for each of the 11 GFTs (see Figure 7.4). In particular, it will identify those two or three GFTs most in need of immediate attention. The software will also analyse trends over time and archive the data from many sites. Tripod-Delta testing sessions usually take place on a quarterly basis, though this can be adjusted to suit the pace of change in a particular work site.

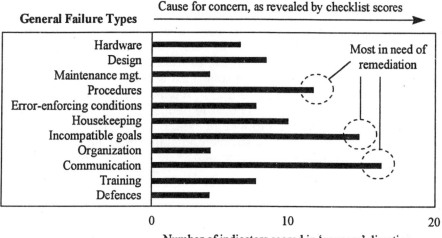

Figure 7.4   A Tripod-Delta Failure State Profile identifying the three GFTs most in need of improvement in the near future (in this case, procedures, incompatible goals and communication)

The key to using Tripod-Delta is to manage the profiles rather than individual indicators. Because indicators are merely symptoms, fixing them individually rarely 'cures' the underlying 'disease'. Line management's task is to review the Failure State Profiles for their respective areas and to plan and implement remedial actions for the two to three 'worst' GFTs obtained on any one testing occasion. In this way, safety management involves the continuous proactive improvement of the underlying causal processes instead of reactive 'fixes' for preventing the last accident.

It should also be noted that only one of the GFTs—defences—is specifically safety-related. The remainder relate to general processes underpinning both quality and safety. As such, a Failure State Profile forms an important part of any line manager's tool box. Safety is not an 'add-on'.

There is a close synergy between Tripod-Delta and the kinds of safety management systems that have been produced by oil exploration and production companies in response to the Cullen Report on the *Piper Alpha* disaster.[9] A safety management system provides the administrative structures necessary to drive good safety practices. It focuses upon the technical and managerial factors associated with hazards. It is top-down and management-led. It is prescriptive and normative—that is, it states how things *ought* to be. It is comprehensive, embracing all hazards and their management requirements. Tripod-Delta, on the other hand, focuses on the organizational and human factors affecting safe working practices. It is bottom-up in its operation. It is created in the workplace using the craft knowledge of task specialists. It is descriptive—that is, it tells how the world *actually is*, rather than how it ought to be. It is deliberately non-comprehensive. It samples a limited number of possible dimensions of safety health. In short, each process supplements and augments the other. The two are intended to work in parallel.

Review and MESH

Although they are applied to different domains, it is appropriate to deal with Review[10] and MESH[11] together, since they are both guided by the same underlying principles and both use the same method of assessment. Both instruments are the natural descendants of Tripod-Delta, although they use ratings rather than indicators to measure the adverse impact of latent conditions. Both were developed at the University of Manchester—in collaboration with British Rail Research and British Airways Engineering, respectively—in the early 1990s and have been used for some years. As indicated in Table 7.1, Review addresses the local and organizational factors affecting human performance in railway operations, while MESH (Managing Engineering Safety Health) assesses the same kinds of issues in the context of aircraft maintenance.

The Tripod-Delta indicators have many obvious merits. They are directly observable and so not easily influenced by the respondent's whims and biases. They have both content validity and face validity. That is, they are directly linked to individual GFTs in a particular operational context and can be seen—by all specialist users—to be relevant to the local task issues (since the indicators were generated by people just like themselves). But they are not without their problems. Creating a database of indicators is a lengthy, expensive and labour-intensive process. Although it is assumed that each item in a related collection of indicators is equally 'diagnostic' of the associated GFT, that is not always the case in practice. Some items can be more discriminating than others; others can suffer from 'floor and ceiling' effects. That is, they would always be answered with a 'yes' or a 'no' and so convey little information.

For these and other reasons, both Review and MESH employ subjective five-point ratings rather than objective 'yes/no' indicators. Many engineers view numbers that cannot easily be given to two decimal places with dark suspicion. What they do not always appreciate, however, is that such impressionistic ratings exploit a natural human talent. We are very good at judging the relative frequencies of particular kinds of events—using an ordinal measurement scale rather than an interval or ratio scale. We encode frequency-of-occurrence data automatically, without conscious effort. When these ordinal estimates of the frequencies of events are compared with the reality (when that is known), the correlation coefficients are often in the region of +0.9 or better.

Clearly defined problems are events like any other and can be estimated in the same way. In both Review and MESH, task specialists—people who actually do the 'hands on' work or first-line managers—are asked to estimate on a scale from 1 (hardly at all) to 5 (very often indeed) how frequently particular kinds of workplace or organizational problems have adversely affected their work over a short, clearly defined time period, or in relation to a few specific jobs.

Another matter that causes concern to those with a background in the 'hard' sciences is the conviction that such ratings are prey to bias. Malcontents, they think, will use the scales as an excuse to complain, while those at the opposite end of the spectrum will say that everything is for the best in the best of all possible worlds, and neither will report the reality. Of course, such people do exist, but the behavioural sciences have been coping with the subjectivity problem for over a century. One answer is to pool data from a large number of people to create an average picture at any one time. In this way, biased views of either kind cancel one another out. Such a method also helps to preserve the anonymity of the respondents (see Chapter 9 for a further discussion of what is needed to achieve a 'minimal blame' or 'just' reporting culture).

Review assesses 16 Railway Problem Factors (RPFs), selected on the basis of extensive field studies. These assessments are made on a regular basis (the intervals vary according to the location) by supervisors in differing activities and locations, via a computer program. The assessors are asked to rate the degree to which each RPF has constituted a problem in their area of work over the last accounting period. The program analyses and archives the data by both location and task. The RPFs are listed below:

- Tools and equipment
- Materials
- Supervision
- Working environment
- Staff attitudes
- Housekeeping
- Contractors
- Design
- Staff communication
- Departmental communication
- Staffing and rostering
- Training
- Planning
- Rules
- Management
- Maintenance.

While railway operations are strung out along thousands of miles of track, aircraft engineering activities are usually located in a centralized cluster of workshops, hangars and offices. This was certainly the case for British Airways Engineering at Heathrow where MESH was developed. As a result, MESH could be configured in a more varied way than Review. Assessments were made of both local and organizational factors, each by different types of personnel. Whereas local factors varied from one workplace to the next (in-line hangars, base maintenance bays, workshops and airworthiness offices and so on), the same eight organizational factors were measured throughout the organization:

- Organizational structure
- People management
- Provision and quality of tools and equipment

- Training and selection
- Commercial and operational pressures
- Planning and scheduling
- Maintenance of buildings and equipment
- Communication.

The ratings of organizational factors are made monthly, and sometimes quarterly, by technical managers—people at the interface between the system as a whole and their particular workplaces. By contrast, assessments of local factors are made by between 20–30 per cent of the 'hands on' workforce in any given location. Assessors are selected randomly and each makes regular assessments, usually on a weekly basis, but sometimes at longer intervals. The assessments are made directly on to a computer. Each assessor logs on anonymously, giving only his or her grade, trade and location. At each assessment the MESH program asks the individual to rate the degree to which each local factor has been a problem with regard to a limited number of specified jobs or within a given work period (for example, a particular day). The assessments are made by moving a mouse-driven cursor along a rating scale. Listed below are the set of local factors created for line maintenance:

- Knowledge, skills and experience
- Morale
- Tools, equipment and parts
- Support (from other sectors)
- Fatigue
- Pressure
- Time of day
- The environment
- Computers
- Paperwork, manuals and procedures
- Inconvenience
- Personnel safety features.

In both Review and MESH (as for Tripod-Delta), these assessments are summarized as bar diagrams, in MESH, for example, each assessor sees first the bar diagram for his or her ratings and then the average ratings for that workplace over the past four weeks. As in Tripod-Delta, the purpose of these profiles is to identify two or three factors most in need of remediation and to track their changes over time.

It should be emphasized that the success of techniques like Tripod-Delta, Review and MESH depends crucially upon the assessors seeing management act on their ratings. The management needs to keep the workforce informed of remedial progress at all times. If nothing is seen to be done, there is no incentive to give assessments and the system dies.

Human Error Assessment and Reduction Technique (HEART)

HEART was developed by Jeremy Williams, a British ergonomist with experience of many hazardous technologies. Its detailed application has been described in detail elsewhere.[12] In this section, we will outline its basic features, describe how error-producing conditions can be ranked according to their relative influences and introduce a new aspect—a principled means of assessing the impact of various violation-producing factors (VPCs) upon a person's likelihood of failing to comply with safe operating procedures.

HEART provides a set of generic task types with their associated nominal error probabilities. They are listed in Table 7.2. The starting point for the HEART analysis is to match the activity to be evaluated to one of the generic tasks listed in Table 7.2. The next, and most important, step (and the one that gives HEART its pre-eminence in this area) is to consult the list of error-producing conditions (EPCs) and to decide which condition(s) are likely to impact upon the performance of the activity in question. Having done this, the nominal error probability is multiplied by a suitably judged proportion of the appropriate EPC factor.

When people are asked to make absolute probability estimates of a particular kind of error type, their judgements may vary by orders of magnitude from person to person. But an extensive survey of the human factors literature has revealed that the effects of various kinds of manipulation upon error rates show a high degree of consistency across a wide variety of experimental situations. The principal error-producing conditions (EPCs) are listed below in rank order of effect. The numbers in parentheses indicate the amount by which the nominal error probability must be multiplied to reflect the influence of each factor:

- Unfamiliarity with a situation that is potentially important, but which is either novel or occurs only infrequently ($\times$ 17)
- Shortage of time for error detection and correction ($\times$ 11)
- Low signal-to-noise ratio—when really poor ($\times$ 10)
- Suppression of feature information that is too accessible ($\times$ 9)
- Absence—or poverty—of spatial and functional information ($\times$ 8)
- Mismatch between designer's and user's model of system ($\times$ 8)
- No obvious means of reversing an unintended action ($\times$ 8)

- A channel capacity overload, particularly one caused by the simultaneous presentation of non-redundant information (× 6)
- A need to unlearn a technique and apply one that requires the application of an opposing philosophy (× 6)
- The need to transfer specific knowledge from one task to another without loss (× 5.5)
- Ambiguity in the required performance standards (× 5)
- A mismatch between real and perceived risk (× 4)
- Poor, ambiguous or ill-matched system feedback (× 4)
- No clear, direct and timely confirmation of an intended action from the portion of the system over which control is to be exerted (× 3)
- Operator inexperience (for example, a newly qualified technician) (× 3)
- An impoverished quality of information conveyed by procedures and person-to-person interaction (× 3)
- Little or no independent checking or testing of output (× 3)
- A conflict between immediate and long-term objectives (× 2.5)
- No diversity of information input for veracity checks (× 2.5)
- A mismatch between the educational achievement level of an individual and the requirements of the task (× 2)
- An incentive to use other, more dangerous procedures (× 2)
- Little opportunity to exercise mind and body outside the immediate confines of the job (× 1.8)
- Unreliable instrumentation that is recognized as such (× 1.6)
- A need for absolute judgements which are beyond the capabilities or experience of an operator (× 1.6)
- Unclear allocation of function and responsibility (× 1.6)
- No obvious way to keep track of progress during task (× 1.4)
- A danger that physical capacities will be exceeded (× 1.4)
- Little or no intrinsic meaning in a task (× 1.4)
- High-level emotional stress (× 1.3)
- Ill-health, especially fever (× 1.2)
- Low workforce morale (× 1.2)
- Inconsistency of meaning of displays and procedures (× 1.15)
- Prolonged inactivity or repetitious cycling (× 1.1 for the first half-hour, and × 1.05 for each hour thereafter)
- Disruption of normal work-sleep cycles (× 1.1)
- Task-pacing caused by intervention of others (× 1.06)
- Additional team members over and above those necessary to perform tasks normally and satisfactorily (× 1.03 per additional person)
- Age of personnel performing perceptual tasks (× 1.02).

Table 7.2   Generic tasks and associated error probabilities (after Williams)

Generic tasks	Nominal error probabilities (5th–95th percentile bounds)
A. Totally unfamiliar, performed at speed with no idea of likely consequence.	0.55 (0.35–0.97)
B. Shift or restore system to a new or original state on a single attempt without supervision or procedures.	0.26 (0.14–0.42)
C. Complex task requiring high level of comprehension and skill.	0.16 (0.12–0.28)
D. Fairly simple task performed rapidly or given scant attention.	0.09 (0.06–0.13)
E. Routine, highly practised, rapid task involving relatively low level of skill.	0.02 (0.007–0.045)
F. Restore or shift system to original or new state following procedures, with some checking.	0.003 (0.0008–0.007)
G. Completely familiar, well designed, highly practised routine task, oft-repeated and performed by well motivated, highly trained individual with time to correct failures but without significant job aids.	0.0004 (0.00008–0.009)
H. Respond correctly to system even when there is an augmented or automated supervisory system providing accurate interpretation of system state.	0.00002 (0.000006–0.00009)
M. Miscellaneous task for which no description can be found.	0.03 0.008–0.11

Table 7.3   Generic violation behaviours and associated nominal probabilities for females

	Generic violation behaviours	Nominal error probabilities for females (5th–95th percentile bounds) × 1.4 for males
A.	Distinctly inconvenient to comply. Potential violator does not feel bound by any implied requirement to comply. Easy to violate. Little likelihood of detection	0.42 (0.28–0.58)
B.	Compliance relatively unimportant. Easy to violate. Little or no inducements to comply.	0.35 (0.20–0.59)
C.	Compliance may be fairly important, but chances of detecting violation low. Personal benefits of violating are high and direct.	0.38 (0.21–0.54)
D.	Personal benefit in violating, though likelihood of detection is moderate to high. Or else compliance fairly important, but chances of detection low.	0.18 (0.11–0.25)
E.	Compliance important, usually legally required, but chances of detection low to moderate.	0.03 (0.007–0.05)
F.	No immediate incentive to violate, but likelihood of violation detection moderate to high	0. 007 (0.001–0.01)
G.	Socially unacceptable, likelihood of detection low and likelihood of unfavourable outcome for violator low.	0.007 (0.003–0.02)
H.	Socially unacceptable, chances of detection high, and chances of bad outcome high.	0.0001 (0.00002–0.003)

Over the past two years, Jerry Williams has compiled a similar list for violation-producing conditions (VPCs), again using an extensive trawl of the psychological and human factors literatures.[13] As before, he created a limited list of generic violation behaviours, each with their associated nominal probabilities for females—unlike errors, there is a large gender effect in non-compliance. For males, the nominal probabilities need to be multiplied by 1.4. The generic situations and their nominal values for females are shown in Table 7.3.

To date, Williams has established the relative impact of eight VPCs. These, together with their weighting factors, are:

- Perceived low likelihood of detection ($\times$ 10)
- Inconvenience ($\times$ 7)
- Apparent authority or status to violate, disregard, or override advice, requests, procedures or instructions ($\times$ 3)
- Copying behaviour ($\times$ 2.1)
- No disapproving authority figure present ($\times$ 2)
- Perceived requirement to obey 'authority figure' ($\times$ 1.8)
- Gender ($\times$ 1.4 for males)
- Group pressure ($\times$ 1.07 per individual encouraging deviation—up to a maximum of five people).

Together, these lists of EPCs and VPCs constitute the best available account of the factors promoting errors and violations within the workplace. The fact that they can be ranked reliably—so that we can assess the relative effects of the different factors—represents a major advance and an extremely valuable addition to the safety-concerned manager's tool box.

The Influence Diagram Approach (IDA)

The Influence Diagram Approach (IDA) has two important aspects. First, it provides the tools for modelling *qualitatively* the influences, existing at various organizational levels, upon adverse outcomes—either in assessing their contributions to past events, or in considering the likelihood of some future event. Second, this qualitative model can also be used to generate *quantitative* measures of the influences of various technical, human and organizational factors on the risks faced by a particular hazardous technology. Once again, these quantitative measures can be derived either reactively in regard to a particular accident or proactively to gauge the probability of some possible adverse event occurring in the future. As such, this approach can play a very informative role in both accident investigation and in the preparation of Formal Safety Analyses. In principle, it is applicable to any kind of hazardous system, though the example of its application given below will be drawn from the maritime domain.[14]

While the particular ways in which bad outcomes can occur may not be easily foreseeable, the varieties of bad outcomes themselves are mostly known for each hazardous domain. In shipping, for example, the ways in which disaster can strike a vessel—or the failure modes—are not likely to extend much beyond fires, collisions, groundings, founderings, piracy and acts of war. The Influence Diagram Approach charts the influences upon a particular failure mode—we will take the example of a vessel grounding on a river bar—at each of a number of levels:

- the *influencing factor* level: this includes the unsafe acts or technical failures immediately responsible for the event,
- the *performance-influencing factor* (PIF) level—these are the immediate workplace conditions that shape the occurrence of human or technical failures,
- the *implementation* level—these are the underlying organizational factors that create the PIFs,
- the *policy* level—this comprises the policy and regulatory factors that determine organizational processes occurring at the implementation level.

Figure 7.5 shows an Influence Diagram relating to the case of a vessel grounding on a river bar. For simplicity, we have combined the PIF and the influencing factor levels—such contractions are allowable within the Influence Diagram methodology. It will also be noticed that the Influence Diagram is entirely consistent with the theoretical framework for the development of organizational accidents set out in earlier chapters of this book.

In practical terms the IDA is performed in stages. First, the Influence Diagram is developed by means of a structured discussion with a group of experts—in this case, marine specialists, though comparable experts would be required for other domains—who have a detailed knowledge of both the operational realities and the broader policy issues. Having identified the factors involved at each level and their respective influences upon the failure mode in question, the next step involves quantifying the influences.

Normally, the business of assigning numbers to human failure probabilities is little more than an art form, but the steps involved in quantifying the elements of the Influence Diagram—and hence deriving the 'top event' likelihood—are relatively simple, workable, elegant and based upon the best available information. The process begins with one of the 'bottom influences' (leaf nodes)—that is, a factor that has no external influences shown as impinging upon it. In

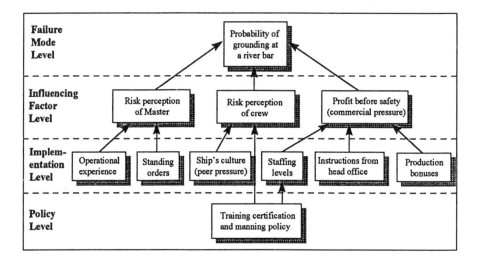

Figure 7.5   A simplified influence diagram showing some of the factors determining the probability of a vessel grounding at a river bar
Normally, the Influence Diagram would have many more elements at each of the levels. The figure has been simplified to correspond with the explanation of the quantification process, given below (after Embrey, 1996)

the case of the simplified example presented in Figure 7.4, this includes all the factors at the implementation level, excluding staffing levels, and the single factor shown at the policy level (normally there are many more). As in the case of creating the qualitative Influence Diagram, the quantification of the individual and combined influences is an iterative process, requiring inputs from the expert assessment team. In order to explain how these calculations are made, we will start with a small part of the influence diagram—the risk perception of the Master and the two factors shown as influencing it: operational experience and standing orders. In the first step the assessment team evaluates the evidence that feedback from operational experience will influence the Master's risk perception either positively or negatively. To help them make this judgement, they are given a graded indicator scale that specifies the nature of the evidence to be taken into account. At the ideal extreme of the scale there could be the statement: 'Results of grounding incidents and near misses are regularly fed back to ships' masters.' The other end of the scale could describe the worst case as follows: 'No feedback from operational experience is available to masters.' The assessment team will need to be guided by an experienced facilitator.

Let us suppose that the evidence before them indicates that feedback is not used effectively to improve the Master's perception of risk. Thus, in answer to the question 'What is the weight of evidence for the availability of feedback from operational experience as an aid to enhancing the Master's risk perception?' the assessment team comes up with an influence ratio (a balance of likelihoods) of 0.20 (good): 0.80 (poor). Let us also suppose that when the same question is asked in relation to standing orders, the team judges the ratio to be 0.20 (used): 0.80 (not used). The next stage is to combine these two influences to derive the likelihood that the Master's risk perception will be accurate or inaccurate. The individual steps involved are shown in Table 7.4.

The steps involves in calculating the unconditional probability of a factor having two or more 'upstream' influences are as follows:

- List the possible combinations of influencing factors.
- The assessment team is asked to evaluate the weight of evidence that the accuracy of the Master's perception will be high or low for each of these combinations of influences. In Table 7.4 it can be seen that the combined effect of poor feedback and unclear standing orders is judged to degrade the accuracy of risk perception more strongly than either factor considered in isolation.
- Each of the conditional assessments is modified by a joint weighting factor obtained from the product of the appropriate individual factor probabilities.
- The overall unconditional probability is derived by multiplying each of the conditional probabilities by the appropriate weighting and then summing over all of these obtained values.

The same basic steps are applied to all the other influences shown in Figure 7.4. The overall probability of grounding at a river bar is obtained by summing the unconditional probabilities of each of three immediate influencing factors: the risk perception of the master; the risk perception of the crew; and the likelihood of profit being placed before safety. Although laborious for the assessment team, the Influence Diagram Approach is capable of yielding qualitative analyses and quantitative risk estimates that are based upon sound theoretical principles and state-of-the-art knowledge engineering techniques.

Table 7.4   The steps involved in calculating the unconditional probability that the Master's risk perception will be inaccurate

If... feedback from operating experience is	and... standing orders are	then...	Weight of evidence that risk perception of the Master is...		Joint weight (feedback × standing orders)
			accurate	inaccurate	
good	clear used	→	0.95	0.05	(0.20 × 0.20) = 0.04
good	not used	→	0.80	0.20	(0.20 × 0.80) = 0.16
poor	used	→	0.15	0.85	(0.80 × 0.20) = 0.16
poor	not used	→	0.10	0.90	(0.80 × 0.80) = 0.64
Unconditional probability (weighted sum) that risk perception of Master is inaccurate is:			**0.25**	**0.74**	

Maintenance Error Decision Aid (MEDA)

The Maintenance Error Decision Aid (MEDA) is a tool, devised by Boeing in collaboration with the FAA and Galaxy Scientific Corporation, for investigating maintenance errors.[15] Although designed for an aviation context, its basic principles are applicable to any safety-critical maintenance activities. The investigation takes place at two levels:

- *Line investigation.* MEDA begins with a paper-based investigation that gives line-level maintenance personnel a standardized way to investigate maintenance errors, their origins and their consequences. It offers front-line engineers a principled means of detecting and removing error-provoking factors at both the workplace and the organizational levels.
- *Organizational trend analysis.* MEDA also provides a means for computerized trend analysis for the maintenance organization.

MEDA is divided into five sections. Sections 1–3 deal with the question 'What happened?'. Section 4 addresses how and why the error occurred. Section 5a pinpoints failed defences and Section 5b outlines potential solutions.

- *Section 1* gathers information about the airline, aircraft type, engine type, time of day, and the like.
- *Section 2* describes the nature of the event (for example, flight delay, cancellation, gate return, inflight shutdown and so on).
- *Section 3* classifies the nature of the lapse. These are broken down into the following categories: improper installation, improper servicing, improper or incomplete repair, improper fault isolation, inspection or testing, foreign object damage, surrounding equipment damage, personal injury. An error is defined as 'the condition resulting from a person's actions, when there is general agreement that the actions should have been other than what they were'.
- *Section 4* takes the form of a contributing factors checklist that must be completed for each of the errors identified in Section 3. These include: information, equipment, tools or parts, aircraft design and configuration, job or task, qualifications and skills, individual performance, environment and facilities, organizational environment, supervision, communication.
- *Section 5a* asks whether there were any current procedures, processes and policies within the system that should have prevented the incident, but which did not.
- *Section 5b* asks what corrective measures have been taken, or

should have been taken at the local level to prevent the recurrence of the incident.

Boeing have distributed MEDA free to all their airline customers. One of its primary aims is to provide a common language to increase communication and cooperation between operators, regulators, manufacturers and the maintenance workforce. A PC-based version has recently been developed for Singapore Airlines Engineering Company, where it is used in conjunction with MESH.

Tripod-Beta

Tripod-Beta is a PC-based tool, created within Shell Exploration and Production, for conducting an incident analysis in parallel with an event investigation.[16] Interaction between these two activities provides the investigators with guidance as to the relevance of their fact-gathering, and also highlights avenues of investigation leading to the identification of contributing latent conditions—General Failure Types in Tripod parlance. Tripod-Beta was born out of a marriage between Tripod theory (discussed earlier in this chapter) and Shell's response to the post-Cullen Report safety case requirements. In Shell, this took the form of the Hazard and Effects Management Process (HEMP). This is designed to provide a structured approach to the analysis of health, safety and environmental hazards throughout the lifecycle of an installation. It is based on four processes: identify, assess, control and recover. Tripod-Beta was designed to be compatible with both Tripod-Delta and HEMP.

Figure 7.6 shows the basic elements of the Tripod-Beta analytical toolkit. As indicated in Chapter 1, an event involves hazards coming into damaging contact with targets (people, assets, environment) as the result of a defensive failure. Starting with the end result, a loss of some kind, the analyst works backwards to determine the nature of the failed defence(s) and the hazard. A short piece of descriptive text is written into each box or node to describe the actual occurrence. The next step is to establish why a particular defence failed. This may be due to either active or latent failures (the boxes below the latent failure nodes with question marks are there to identify which of the 11 GFTs were implicated—see page 132, this chapter). In the case of an active failure, there are likely to be preconditions within the workplace. These, like active failures, are identified by short pieces of descriptive text. Preconditions are likely to be the product of identifiable latent failures. Once again, they are described and the associated GFTs identified. The procedure continues until all the identifiable factors have been identified and described. As the ana-

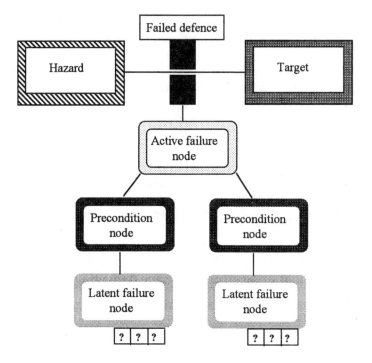

Figure 7.6   The basic units of the Tripod-Beta event analysis

lyst progresses, he or she will link the causally related nodes and boxes by lines on the computer screen.

Thus, the Tripod-Delta software provides the means to assemble the facts learned in the investigation and to manipulate them on-screen to produce a graphic picture of the event and its causes—an incident tree. The program also checks whether the logic of the tree structure (its labelling and connections) conforms to the Tripod theory and HEMP. Once anomalies have been resolved, a draft accident report is automatically generated for final editing on a word-processing package.

Summary of the Main Principles of Error Management

1   The best people can sometimes make the worst errors.
2   Shortlived mental states—preoccupation, distraction, forgetfulness, inattention—are the last and the least manageable part of an error sequence.
3   We cannot change the human condition. People will always make errors and commit violations. But we can change the conditions

under which they work to make these unsafe acts less likely.

4   Blaming people for their errors—though emotionally satisfying— will have little or no effect on their future fallibility.

5   Errors are largely unintentional. It is very difficult for management to control what people did not intend to do in the first place.

6   Errors arise from informational problems. They are best tackled by improving the available information—either in the person's head or in the workplace.

7   Violations, however, are social and motivational problems. They are best addressed by changing people's norms, beliefs, attitudes and culture, on the one hand, and by improving the credibility, applicability, availability and accuracy of the procedures, on the other.

8   Violations act in two ways. First, they make it more likely that the violators will commit subsequent errors and, second, it is also more likely that these errors will have damaging consequences.

Notes

1   J. Reason, *Human Error*, (New York: Cambridge University Press, 1990).
2   K. Ishikawa, *Guide to Quality Control*, (Tokyo: Asian Productivity Organization, 1976).
3   S.T. Fiske and S.E. Taylor, *Social Cognition*, (Reading, MA: Addison-Wesley, 1978).
4   H.M. Lefcourt, 'The function of the illusions of control and freedom', *American Psychologist*, May 1973, pp. 417–25.
5   D. Meister, 'Human error in man-machine systems', in S. Brown and J. Martin (eds), *Human Aspects of Man-Made Systems*, (Milton Keynes: The Open University Press, 1977, pp. 299–324). See also the paper by S.C. Brown in the same volume (pp. 44–53).
6   E. Hollnagel, *Human Reliability Analysis: Context and Control*, (London: Academic Press, 1993). Center for Chemical Process Safety. *Guidelines for Preventing Human Error in Process Safety*, (New York: Center for Chemical Process Safety of the American Institute of Chemical Engineers, 1994). B. Kirwan, A *Guide to Practical Human Reliability Assessment*, (London: Taylor & Francis, 1994). A.I. Glendon and E.F. McKenna, *Human Safety and Risk Management*, (London: Chapman & Hall, 1995).
7   E. Wiener, B. Kanki and R. Helmreich, *Crew Resource Management*, (New York: Academic Press, 1993). N. Johnston, N. McDonald and R. Fuller, *Aviation Psychology in Practice*, (Aldershot: Avebury, 1994).
8   P. Hudson, J. Reason, W. Wagenaar, P. Bentley, M. Primrose and J. Visser, 'Tripod-Delta: proactive approach to enhanced safety', *Journal of Petroleum Technology*, 40, 1994, pp. 58–62.
9   The Hon. Lord Cullen, *Public Inquiry into the Piper Alpha Disaster*, (Department of Energy, London: HMSO, 1990).
10   J. Reason, *Review. I. Overview. II. Theory*, (Derby: British Railways Board, 1993). See also, S. Tozer, *Review: Summary of Pilot Study*, (London: Railtrack Safety &

Standards Directorate, 1994). It should be noted that Railtrack have recently changed the name of Review to Target.

11   J. Reason, *Comprehensive Error Management in Aircraft Engineering: A Manager's Guide*, (London Heathrow: British Airways Engineering, 1995).

12   J.C. Williams, 'HEART: a proposed method for assessing and reducing human error' in *Proceedings of the Ninth Advances in Reliability Technology Symposium*, (Bradford: University of Bradford, 1986). J.C. Williams, 'A data-based method for assessing and reducing human error to improve operational performance', in *Proceedings of IEE Fourth Conference on Human Factors in Power Plants*, Monterey, CA, 6–9 June 1988.

13   J.C. Williams, *Assessing the Likelihood of Violation Behaviour: A Preliminary Investigation*, (Manchester: Department of Psychology, University of Manchester, 1996).

14   D. Embrey, 'Incorporating Management and Organizational Factors into Probabilistic Safety Assessment', *Reliability Engineering*, 38, 1992, pp. 199–208.

15   Boeing, *Maintenance Error Decision Aid*, (Seattle: Boeing Commercial Airplane Group, 1994).

16   Tripod-Beta was designed by Dr Miles Benson, an independent human factors and software consultant. He also wrote the software for Tripod-Delta, Review and MESH. See also J.A. Doran and G.C. van der Graaf, 'Tripod-Beta: incident investigation and analysis', *Proceedings of the International Conference on Health, Safety & Environment (Society of Petroleum Engineers)*, New Orleans, LA, 9–12 June, 1996.

8 The Regulator's Unhappy Lot

Regulators in the Frame

The regulators' lot—like the policeman's—is not a happy one. Not only are they rarely loved by those they regulate, they are now ever more likely to be blamed for organizational accidents. Over the past 30 years, the search for the causes of a major catastrophe has spread steadily outwards in scope and backwards in time to uncover increasingly more remote contributions. Prominently and frequently featured in this extended causal 'fallout', are the decisions and actions of the regulatory authority. This chapter examines the difficult and complex role of the regulator in limiting the occurrence of organizational accidents.

First, we will look briefly at five tragic events in which regulatory failures—or, at least, shortcomings in the regulatory process—were implicated. We do this in order to gain a better understanding of how regulators (or the regulatory process) can contribute to the breakdown of complex well defended technologies. The five case studies are:

- the *Challenger* spacecraft explosion (28 January 1986)
- the King's Cross Underground fire (18 November 1987)
- the *Piper Alpha* platform explosion (6 July 1988)
- the Fokker-28 crash at Dryden, Ontario (10 March 1989)
- the Piper Chieftain crash at Young, New South Wales (11 June 1993).

Regulated Accidents

Challenger: When Deviance Becomes the Norm

The Presidential Commission, investigating the loss of the *Challenger* space shuttle and its seven astronauts, concluded that the explosion was caused by a faulty seal, or O-ring, on one of the rocket boosters.[1] This allowed flames to escape that ignited an external fuel tank. The Commission's report was published in June 1987. Since then, the *Challenger* accident has been the subject of an intensive study by Diane Vaughan, a Boston College sociologist. This work has recently been published and represents one of the most detailed and compelling analyses yet made of an organizational accident.[2] It also challenges the widely held view that the accident happened because NASA and its prime contractor, Morton Thiokol, did not do their jobs properly.' The impression conveyed by the Presidential Commission report and press accounts is that 'flawed decisions' were made by the middle managers present at the dramatic eve-of-launch teleconference. The received wisdom was that these managers, driven by production pressures and political concerns about the agency, turned a blind eye to the O-ring defects, violated safety rules and went ahead with the launch in order to meet the programme deadlines. These analyses focused on individual failures. As Diane Vaughan expressed it, they '... conveyed an imagery of evil managers, so that the incident appeared to be an anomaly: a peculiarity of the individuals who were in responsible decision making positions at the time.'

Diane Vaughan's own conclusions were quite the reverse. She argued that the accident happened because those involved in the launch decision did precisely what they were supposed to do.

> It can truly be said that the Challenger launch decision was a rule-based decision. But the cultural understandings, rules, procedures and norms that had always worked in the past did not work this time. It was not amorally calculating managers violating rules that was responsible for the tragedy. It was conformity.

Moreover, she found that the *Challenger* tragedy was not an anomaly peculiar to NASA. It was shaped by factors common to many other organizations.

What drove NASA inexorably towards the tragic outcome was the insidious erosion of the standards against which they regulated themselves. While the Presidential Commission was shocked by NASA's frequent use of the phrase 'acceptable risk', Vaughan's analysis reveals that flying with 'acceptable risks' was an integral part of NASA's culture—or, indeed, of any aviation culture. In fact, the 'acceptable risks' on the space shuttle filled six volumes.

> At NASA, problems were the norm. The word 'anomaly' was part of everyday talk ... . The whole shuttle system operated on the assumption that deviation could be controlled but not eliminated.

In short, NASA had created a closed culture that, in Diane Vaughan's words, 'normalized deviance'. What seemed to the outside world like reckless risk-taking was to NASA managers a reasonable and sensible way of doing their jobs. The worrying fact about this and other organizational accidents is that it is very difficult to identify exactly when things started to go wrong since each step and each decision proceeded naturally from the last.

No fundamental decision was made at NASA to do evil. Rather, a series of seemingly harmless decisions were made that incrementally moved the space agency towards a catastrophic outcome.

Three safety units oversaw the Shuttle operations.[3] The most significant of these was NASA's internal body—the Safety, Reliability and Quality Assurance Program (SR & QA). The Presidential Commission identified three principal failures in the SR & Q's monitoring and surveillance responsibilities:

1   They failed to discover and rectify the confusion that surrounded the assignment of criticality categories to Shuttle components according to the seriousness of their failure consequences. The solid rocket booster joint was originally listed as C1-R. The 'C1' was the highest criticality rating, indicating 'loss of life or vehicle', but the 'R' meant that there were redundancies, or back-ups to its possible failure. Later, the criticality rating was changed to the more serious C1 designation. But after the accident it was discovered that many internal documents still listed the joint as C1-R, leading some NASA managers (and even some SR & QA staff) to believe that there was defensive redundancy.
2   The SR & QA staff failed to compile and disseminate trend data relating to in-flight O-ring erosion. Between the tenth and the twenty-fifth mission (the *Challenger* flight), more than half the flights experienced O-ring problems. Had these data been circulated, NASA administrators would have had essential information regarding the history and scope of the joint weaknesses.
3   The Commission found three problem-reporting failures. First, the SR & QA did not establish an adequate means for reporting Shuttle problems up the NASA hierarchy. Second, they failed to create a precise set of requirements for reporting inflight anomalies. Third, they failed to detect violations of problem-reporting requirements.

In addition, no member of the SR & QA staff was present at the teleconference between Marshall Space Flight Center and Morton Thiokol engineers on the night before the launch. On this occasion,

Morton Thiokol engineers expressed considerable disquiet about launching after such an unusually cold night. Instead of being asked to prove that the launch was safe, however, the Morton Thiokol engineers were challenged to prove that it was unsafe. The absence of any SR & Q representatives from this critical decision-making meant that another opportunity to express safety concerns was missed. The reason for their absence? No one thought to invite them.

A theme that links all of the regulatory problems to be discussed here is lack of resources. Between 1970 and the *Challenger* disaster, NASA reduced its safety and quality control personnel by 71 per cent. The SR & QA staff at Marshall Space Center, with oversight responsibility for the rocket booster project, had been cut from around 130 to 84. Overall, safety, reliability and quality staff comprised about 2 per cent of NASA's 22 000 staff. Both the director and deputy director of the SR & QA unit had other duties, so that each spent around 10 and 25 per cent of their time on the Shuttle programme, leaving them very little time for safety issues.

King's Cross Underground Station: A Blinkered View of the Law

Just after the evening rush hour, a lighted match or cigarette passed through a crack in a wooden escalator and set light to an accumulated heap of greasy fluff and waste beneath.[4] Twenty minutes later, flames shot up the escalator shaft and hit the ceiling of the ticket hall, killing 31 people.

Desmond Fennell, the Queen's Counsel heading the subsequent Inquiry, devoted a chapter of his report to the role of the Railway Inspectorate, the regulator. He begins by asserting that the Railway Inspectorate took an overly restricted view of their responsibilities.

> In my view the Railway Inspectorate was mistaken in its interpretation of the law in believing, if London Underground discharged its duty to have due regard to the safety of operations, it had discharged all its statutory duties for the health and safety of passengers .... Even making allowances for the Railway Inspectorate's misunderstanding of their responsibilities [under the law], it is my view that the level of resources and degree of vigour they applied to enforcement ... were insufficient. It was in this climate that poor housekeeping and potentially dangerous conditions in underground stations were allowed to persist.

In the years preceding the disaster, fires on London Underground had been commonplace, though usually of a very minor nature—cardboard cartons and the like. They were given the comforting label of 'smoulderings'. Confident that these apparently trivial events were being adequately covered by the routine inspections of the London

Fire Brigade, the Railway Inspectorate decided, three years before the King's Cross disaster, that it no longer wanted to receive copies of such reports from the Fire Brigade. As Fennell put it, the Chief Inspector '... [now] conceded that this was an unfortunate decision'.

Although the Inspectorate believed the primary responsibility for fire hazards lay with the London Fire Brigade, it had responded to two escalator fires in 1973 by writing to the Chief Operating Manager of the (then) London Transport Executive suggesting a drive to clear away accumulations of inflammable rubbish under escalators. However, no mention was made of risk to passengers. Neither the Chief Inspector nor his staff had '... ever conceived the possibility of an escalator fire rapidly developing and endangering life'. Moreover, he could be excused for this viewpoint on two grounds. First, there was no obvious precedent—the 'flashover' that had wreaked such havoc in the ticket hall was a relatively new phenomenon. Second, as in other railway operations, regulators—not unreasonably—saw the main risks to passengers as being associated with moving trains rather than static stations.

Fennell concluded his account of the Railway Inspectorate's role in the disaster by observing that it had made insufficient use of its powers and had not devoted enough attention to London Underground '... to create the tension necessary to ensure safety'.[5] He was also critical of the Inspectorate's 'cosy' relationship with London Underground and its lack of vigour in checking upon the implementation of agreed safety improvements. We will return to this issue of regulatory style at later points in this chapter.

Piper Alpha: Superficial Inspections and Incomplete Sampling

The Public Inquiry into the *Piper Alpha* disaster (see Chapter 5 for a brief description of the event) was headed by Lord Cullen, a senior Scottish judge.[6] His report was published in two volumes. The first dealt with the background, nature and aftermath of the disaster. A chapter is devoted to the regulatory deficiencies—in particular, the stark contrast between the findings of the Department of Energy's inspections and what was revealed by the Inquiry. The second volume looked to the future and outlined a new safety regime that has subsequently revolutionized the regulation and management of safety in offshore installations.

The *Piper Alpha* platform received three visits from an inspector in the Department of Energy's Safety Directorate in the year preceding the disaster. The first was a routine inspection that identified a number of minor points requiring no immediate follow-up. The second followed a fatal accident and identified weaknesses in shift handovers and the permit-to-work system. The third took place 10 days before the disaster and concentrated on areas in which construction work was in progress.

It was the report on this last visit that contrasted so starkly with the Inquiry findings. In particular, it accepted that Occidental had 'tidied up' the weaknesses in the permit-to-work scheme and shift handovers, when deficiencies in these areas were subsequently found to be major contributors to the disaster. A checklist existed to facilitate the evaluation of permit-to-work systems, but the inspector did not have it, nor was he experienced in the use of such procedures. He failed to inspect Occidental's Operating Procedures manual in connection with the permit-to-work system because he felt he did not have the time to carry out a full audit, requiring two to three days. In the event, he was only able to devote 10 hours to his inspection.

The inspector was also unaware of the practice of switching the diesel fire pumps to manual mode during shifts in which diving took place. Nor did he inspect the deluge system or establish the frequency with which lifeboat drills were carried out. In short, he failed to spot many of the factors that contributed to the disaster a few days later.

When these discrepancies were pointed out to the Director of Safety at the Inquiry, he said, with considerable justification, 'I think that within the context of carrying out an inspection and the very wide-ranging Inquiry that is going on here, there is a total difference in approach'.[7] The Director of Safety described the nature of inspections as follows:

[An inspection is] essentially a sampling exercise. The inspector samples and audits the state of the equipment and management procedures. He talks to personnel and seeks to obtain an over-all picture of how well the installation is being operated, maintained and managed. An inspector must exercise his professional judgement in determining the scope and depth of the inspection and is selected, trained and supervised by line management to this end. He is not given a fixed list of procedures, equipment and items which he must tick off in the form of a check list. This could create considerable difficulties given the variety of operations, working procedures and installations involved. In addition, it would lead to operators anticipating those areas in which an inspector always checked.

Given the natural constraints of the inspection process, there are many who would have considerable sympathy with this approach. Lord Cullen, however, did not. He gave the following judgement on the regulatory process.[8]

Even after making allowance for the fact that the inspections were based on sampling, it was clear to me that they were superficial to the point of being of little use as a test of safety on the platform. They did not reveal a number of clear cut and readily ascertainable deficiencies. While the effectiveness of the inspections has been affected by persistent undermanning and inadequate guidance, the evidence led me to question, in a fundamental sense, whether the type of inspection practised by the Department of Energy could be an effective means of assessing or monitoring the management of safety by operators.

Lord Cullen then devoted a large part of the second volume of the Inquiry report to recommending a new mode of regulation, which has subsequently been implemented. This is discussed later in this chapter.

The Fokker-28 Crash at Dryden: The Safety Net that Failed

An Air Ontario Fokker-28, with 65 passengers and four crew, took off from Dryden Municipal Airport just after midday on 10 March 1989. After takeoff, the aircraft failed to gain altitude and crashed just under one kilometre beyond the end of the runway. Twenty-one passengers and three crew members died. Freezing rain had been forecast and there was a heavy snowstorm in progress at the time of takeoff. The aircraft had not been de-iced and the immediate cause of the crash was surface contamination on the wings.

On the face of it, this might appear to be an open-and-shut case of pilot error since the captain was obviously mistaken in deciding to proceed with the takeoff under those conditions. However, the Inquiry conducted by Commissioner Virgil Moshansky, an Alberta judge, took two years, received 34 000 pages of testimony and heard 166 witnesses to come to the following conclusion:[9]

> The accident at Dryden ... was not the result of one cause, but of a combination of several related factors. Had the system operated effectively, each of these factors might have been identified and corrected before it took on significance ... the accident was the result of a failure in the air transportation system [as a whole].

Featuring prominently among these causal factors were the shortcomings identified at many different levels of the regulator, Transport Canada. This agency was not in a happy position at the time of the crash, being caught in a classic double bind. Deregulation had resulted in a greatly increased workload. At the same time, spending cuts imposed by the Canadian government meant that the number of people available and qualified to carry out these extended responsibilities was very much reduced. Transport Canada was also trapped within a bureaucratic nightmare: for example, approval of budget increases required separate cases to be submitted to 11 levels of the appropriate Ministry. Safety initiatives had been consistently down-graded or denied for several years. The result of these constraints was that the air carrier inspection and monitoring safety net failed. Several factors relating to Transport Canada were identified by the Moshansky Inquiry as contributing to the Dryden crash. These included:

- lack of guidance to air carriers on the need for de-icing,
- no audit of F-28 operations,
- no approved Airplane Operating Manual or Flight Operations Manual,
- no Minimum Equipment List six months after the aircraft had entered service,
- incomplete, conflicting and ambiguous air navigation orders (the operating regulations),
- Dispatcher training and certification entirely in the hands of the air carriers—the dispatcher responsible for the F-28 flights on that day was underqualified, prepared an incorrect flight release, did not notify the crew of freezing rain at Dryden, and failed to advise the crew that an overflight of Dryden was indicated.

The Inquiry Report summarized Transport Canada's involvement in the accident sequence as follows:[10]

Transport Canada, as the regulator, had a duty to prevent the serious operational deficiencies in the F-28 programme ... . Had the regulator been more diligent in scrutinising the F-28 implementation at Air Ontario, many of the operational deficiencies that had a bearing on the crash of flight 1363 could have been avoided.

There can be little doubt that these regulatory failings arose from deep-rooted systemic failures rather than from the inadequate performance of individuals at the 'sharp end'. The source problems lay high up in Transport Canada and in the financial climate imposed by the government of the time. The operational regulators who testified at the Inquiry were very forthright in their condemnation of both the existing regulations and the 'chronic inaction' of Transport Canada's senior management. The Inquiry also acknowledged[11] that, for the most part, Transport Canada was staffed by 'competent and dedicated persons who are sincerely doing their best to ensure a safe air transportation system for the public, at times under trying and frustrating circumstances'.

Let us give Commissioner Moshansky the last word on the Dryden tragedy:[12] 'After more than two years of intensive investigation and public hearings, I believe that this accident did not just happen by chance—it was allowed to happen.' In this instance, the failure of the regulatory safety net was but one part, albeit a significant one, of the malaise afflicting the entire Canadian air transport system.

Piper Chieftain crash at Young, NSW: Conflicting Goals

At about 1918 hours on 11 June 1993, a small commuter aircraft flying out of Sydney, with five passengers and two pilots onboard, struck trees on a small hill on the approach to the aerodrome at Young, New South Wales (NSW), and crashed. Six people were killed by the impact or the subsequent fire, and one survivor died in hospital the next morning. By the standards of aviation accidents, this was a relatively small disaster, but it was to have far-reaching consequences—particularly for the regulator, the Australian Civil Aviation Authority (ACAA).[13]

The investigation was conducted by the Bureau of Air Safety Investigation (BASI), an agency that had recently adopted the accident causation model outlined in Chapter 1 as a framework for uncovering organizational and systemic contributions.[14] In addition to identifying failed or absent defences, active failures and local error-provoking factors, the model also directs the investigator to consider organizational factors and latent failures. Among these, a number of regulatory contributions were identified by the BASI accident report. These included the following factors:

- *Conflicting goals.* The activities of the ACAA's Safety and Standards Division appeared to be biased towards promoting the viability of the operator (Monarch Air—a small commuter airline in commercial difficulties) rather than serving the safety needs of the travelling public.
- *Poor division of responsibilities.* Both responsibilities and crucial information were shared between the ACAA's Head Office in Canberra and the local district offices—that included the flight operations inspectors who actually carried out the required surveillance. This meant that 'at a regional level, no single person had a comprehensive overview of the safety health of Monarch'.
- *Poor planning.* The district office responsible for Monarch neither formulated nor undertook an effective programme of operational surveillance of Monarch. Because the assigned flight operations inspector had a heavy workload his approach was reactive rather than proactive. His checking on Monarch was therefore largely confined to document inspections. BASI estimated that, for the number of hours the aircraft had flown in

the year prior to the crash, there should have been a minimum of 14 hours and an average of 17 hours en route surveillance. In reality, there was none. This meant that the regulator was unaware of the actual standard of Monarch's inflight operations. One indicator that could have been significant in Monarch's case was its financial circumstances, but no mechanism existed for the ACAA to consider any possible connection between an organization's financial state and its ability to fly safely.

- *Inadequate resources.* The local flight operations inspector was overloaded and underresourced. He was responsible for the surveillance of around 40 companies holding Air Operators Certificates (AOC). In addition, he was required to take over his manager's duties for lengthy periods while still supervising the AOC holders assigned to him. BASI assessed his workload as being unachievable in full.
- *Ineffective communications.* Effective regulation requires a good understanding of the supervised organization, but communications between Monarch and the ACAA district office were carried out at a largely formal level with no indication of a close working relationship between the two parties. There was little personal contact between the assigned inspector and the individuals responsible for Monarch flight operations.
- *Poor control.* The investigators found no evidence of effective control of the Monarch surveillance programme by the ACAA district office. Furthermore, the assigned inspector lacked adequate support and supervision from his line managers.
- *Poor operating procedures.* The Safety Regulation and Standards Division lacked the procedures to ensure that Monarch continued to meet the standards required of an AOC holder. Although deficiencies in Monarch's flight operations had been identified a few months before the accident, the ACAA appeared to be reluctant to take decisive action to improve Monarch's operating standards. In the absence of a coordinated monitoring strategy, a number of meetings with Monarch had failed to achieve the significant improvements required.

The Monarch investigation received a good deal of attention from the Australian media, and widespread public concern was expressed about the safety of small commuter airlines and the effectiveness of the ACAA's surveillance and monitoring programme. In May, 1994, this concern came to a head when the Opposition Transport spokesman in the Federal Parliament made strong allegations about safety violations by Seaview Air, a small charter airline operating between Lord Howe Island and the mainland. The source of these charges was the former Chief Pilot of Seaview Air. They were vigorously denied by the company's proprietor as being '... unfounded, scurrilous, defamatory and quite incorrect'. In this, Seaview's owner was strongly supported on the same day by the ACAA's director of air safety.[15]

In October, 1994, an Aero Commander owned by Seaview Air, crashed into the sea between Newcastle, NSW, and Lord Howe Island causing the deaths of eight passengers and the pilot. Just after the crash, the ACAA's director of air safety regulation was quoted by the *Sydney Morning Herald* as claiming 'We have been very actively looking at how Seaview have been conducting their operation and we consider their operation quite satisfactory'. The same article went on to reveal that, in January 1994, the ACAA had warned Seaview Air that the airline risked being grounded unless 'serious operational deficiencies' were remedied. The ACAA's letter described these problems as 'a major cause for concern from a safety and regulatory compliance point of view'. These deficiencies included:

- Overloading of the aircraft being 'the norm for some time.'
- Flying without a back-up navigation aid.
- Advertising the airline as a regular passenger service when in fact it was a charter service.
- Training deficiencies in the case of some pilots.
- A pilot flying without a flight plan.

The editorial in the *Sydney Morning Herald* on 7 October 1994, six days after the Seaview tragedy, reported that the ACAA had altered its earlier position and conceded that it had indeed told Seaview that it was violating the safety rules, but these violations had been satisfactorily rectified. The same editorial also observed that, when the matter had been raised in the Federal Parliament in May, Mr Brereton, the Minister for Transport, had merely referred the Opposition questions to a parliamentary committee. The editorial commented 'This was essentially a housekeeping response when, in the light of subsequent events, a more direct shake-up might have been required'.

The shake-up was not long in coming. One week after the Seaview disaster, Mr Brereton sacked the ACAA's head of safety. Twelve days after the crash, the Federal police were called in to investigate claims of corruption in the ACAA, stemming from the Seaview accident.[16] In the same week, Mr Brereton announced that the Federal Cabinet had agreed to establish and fund a new independent Air Safety Agency, replacing the ACAA's Department of Air Safety Regulation.[17] He felt that this was the most appropriate way of addressing 'the inherent conflict between the ACAA's commercial and policing functions'. He also announced that government funding for safety regulation was to be increased by over 20 per cent. Meanwhile, Mr Brereton struggled to resist calls for his resignation over his handling of the ACAA (the Opposition called him 'Pontius Pilate in the snow' when it was confirmed that he had continued his skiing holiday after hearing of the Seaview tragedy). In the following year, the Civil Aviation Authority was disbanded and replaced by two agencies: the Civil Aviation Safety Authority and Air Services Australia, the former to regulate air safety and the latter to cover air traffic control and similar aspects of the old ACAA's business. How this arrangement will fare after the recent change of government has yet to be discovered.

Deregulation of the aviation industry in Australia had produced a commuter sector, or third-tier of airlines, comprising a large number of single-pilot, single-plane operators ferrying up to a dozen passengers in and out of Australia's regional centres. Most of them operated under the relatively lax rules governing charter operations. Unlike the United States, where new operators must establish that they have sufficient resources to maintain their aircraft, it was the case in Australia that 'those who could beg, borrow or steal a decent enough plane can become their very own airline magnate'.[18]

This might have been a manageable situation had the ACAA the resources and the overriding aim of maintaining aviation safety, but it was an enterprise agency that was financed, in part, by those it regulated. As a result, the ACAA was quite naturally concerned with sustaining the commercial viability of its client operators in the commuter sector. It was this goal conflict that eventually—and probably inevitably—brought about its demise.

US Regulators under Fire

The following headline appeared in the *New York Times* in November, 1995: 'FAA's Lax Inspection Setup Heightens Dangers in the Skies.' The following feature article by Adam Bryant chronicled some of the problems besetting the Federal Aviation Administration (FAA), one of the world's largest and most prestigious regulatory bodies.

An investigation by the *New York Times*—based on government documents and interviews with inspectors, agency officials and industry experts—identified what the *Times* regarded as two major shortcomings in the FAA's inspection system.[19] First, inspectors did not appear to be held accountable for not discovering or pursuing deficiencies that were later implicated in fatal accidents. The FAA's failure to deal aggressively with lax oversight was identified by the National Transport Safety Board (NTSB), the US air accident investigator, as a factor in three fatal airline crashes as well as an accident involving one of the FAA's own aircraft. Second, a significant number of inspectors still lacked training on the planes and equipment they were required to oversee. For example, many inspectors said they were assigned to overseeing a fleet of Boeing 737s without ever receiving training on the aircraft. In the year 1994–95, only a handful of the 143 inspectors who oversaw pilots flying a widely used turbo-prop plane, the ATR, were fully qualified to fly that type of plane themselves. This lack of regulator training was cited by the NTSB as a factor in three airline accidents since 1983.

In response to these criticisms, agency officials pointed to the nation's air accident rate—among the lowest in the world—as proof that the system worked. But the critics countered with the assertion that resting on a safety record measured by the low rate of bad outcomes is a mistake akin to regarding a bald tyre as safe before it blows out. Senator William S. Cohen, Chairman of the Congressional committee investigating the FAA's inspection system was troubled by the FAA's unwillingness to acknowledge the holes in its safety net. 'Deny, defend and deflect,' he said, 'That had been the attitude of the FAA.'

Like many comparable agencies in other countries, the FAA inspectorate is not overresourced. It has 2500 safety inspectors to oversee 7298 airline aircraft, 184 434 general aviation aircraft, 682 959 active pilots, 4817 repair stations, 656 pilot training schools and 192 maintenance schools. Each year, an inspector will perform more than 400 inspections, any one of them taking between five minutes to five weeks. Their average salary is $57 500.

As elsewhere, the deregulation of the airline industry in 1978 has not made the regulators' work any easier. Previously, they felt that they could trust workers in the industry to do a good job. Now, however, with one in four small airlines going out of business each year and new ones taking their place, there are strong commercial pressures to cut safety corners. As one experienced inspector put it, 'You have mechanics that falsify documents and overlook things that are obvious to them because they are afraid of reprisals from their bosses. The trust is gone'.

The FAA is not the only US regulatory agency under attack. On 4 March 1996, *Time* featured a major article on the Nuclear Regulatory Commission (NRC), the body responsible for overseeing the safety of 110 commercial reactors in the United States. Its headline ran 'Two gutsy engineers in Connecticut have caught the Nuclear Regulatory Commission at a dangerous game: routinely waiving safety rules to let plants keep costs down and stay online'.[20]

The problem related to the use of a deep body of water, called the spent-fuel pool.[21] Every 18 months, the type of nuclear power plant in question was shut down so that the fuel rods can be replaced. The old rods—radioactive and very hot—were moved into racks within the pool. Because the Federal Government has not created a storage site for high-level radioactive waste, these pools have become *de facto* nuclear dumps with many filled to near-capacity. If the system failed, the pool could turn into a bubbling cauldron billowing radioactive steam. Furthermore, if the pool were to be drained by an earthquake, technical failure or human error, there would be a meltdown of multiple cores occurring outside the containment area, releasing massive amounts of radioactive material and rendering large tracts of the United States uninhabitable.

To control these risks, Federal regulations required that older plants, without up-to-date cooling systems, move only one-third of the rods into the pool. In the plant in question, one of its senior engineers discovered that all the hot fuel was being dumped into the pool at once. It also turned out that this practice had been going on for at least 20 years. Instead of allowing the required 250-hour cooling down period, the fuel was being moved just 65 hours after shutdown. By this device, the plant reduced the downtime for each refuelling by two weeks, saving around seven million dollars for the replacement power costs.

Thus began a three-year struggle to put matters right. For 18 months the engineer's supervisors denied that the problem existed and refused to report it to the NRC. Eventually, the management called in outside consultants to prove the engineer wrong, but they ended up by agreeing with him. Finally, the engineer took the issue directly to the NRC and discovered that the nuclear regulators had known about this practice for at least 10 years without acting to halt it. The NRC argued that the practice was widespread and was safe so long as the plant's cooling system was designed to handle the heat load—although this was not the case for the plant in question. It was also discovered that plants in three other states had similar fuel-pool problems.

In a written response to faxed questions from *Time*, the executive director of the NRC stated that 'The responsibility for safety rests with the industry. ... The NRC is essentially an auditing agency'. With just four inspectors for every three plants, '... we have to focus on the issues with the greatest safety significance. We can miss things'.

As elsewhere, the NRC is under attack from all sides. Delaware Senator, Joseph Biden, who is pushing to create an independent nuclear safety board outside the NRC described the NRC's regulatory style as that of 'the fox guarding the hen house'. Critics of the NRC also point out that the US nuclear industry can veto the appointment of commission nominees whom it regards as too hostile, while regulators 'enjoy a revolving door to good jobs' at the nuclear companies. The executive director of a whistle-blower support group called 'We the People' commented, 'It all comes down to money... .

When a safety issue is too expensive for the industry, the NRC pencils it away.' Meanwhile, the nuclear industry itself has complained that many NRC rules boost costs without improving safety because 'the regulatory system hasn't kept pace with advances in technology'.[22] Like its counterparts in other countries, this regulatory body finds itself between a rock and a hard place.

Damned if They Do and Damned if They Don't

Regulatory bodies worldwide seem to be hopelessly trapped in a mesh of double binds. Consider the following:

- Workload has increased as resources have been slashed.
- Regulators are regularly accused of lax oversight and overly collusive relationships with their clients, while the clients themselves often regard the regulators as intrusive, obstructive, threatening, rigid, out-of-date, ignorant and generally unsympathetic to their commercial pressures.
- Accident inquiries find regulators guilty of not being fully acquainted with all the details of their clients' operations and of missing important contributing factors, but the only means they have of obtaining this information is from the operators themselves or from periodic inspections and follow-ups. After an accident, these omissions take on a sinister significance, but for regulators, armed only with foresight, they are but one of many possible contributions to a future accident. As stated earlier, warnings are only truly warnings if we know what kind of an event the organization will suffer.
- Front-line regulators are generally technical specialists, yet major accidents arise from the unforeseen—and often unforeseeable—interactions of human and organizational factors whose role is only now being acknowledged by health and safety legislators, and then in the most general terms.

In short, regulators are in an impossible position. They are being asked to prevent organizational accidents in high-technology domains when the aetiology of these rare and complex events is still little understood—as this book has tried to make plain.

But it was not always the case. Some of the most dramatic reductions in accident rates—usually involving individuals facing clearly defined hazards in particular situations—have been brought about by the introduction of safety-related legislation combined with effective regulation and enforcement. We will look briefly at some of these success stories in order to redress the balance.

Legislation and Regulation: Some Major Successes

After one of General Grant's battles in the American Civil War, a Washington politician, unhappy at the high losses (around 19 000) suffered by the Union forces, accused Grant of killing more of his soldiers than the railroads. Although, at first sight, this appears to be little more than political rhetoric, a closer look at the number of employees killed in nineteenth-century railroad operations indicates that this was hardly an exaggeration. In the year 1892, for example, there were 821 415 railroad employees in the United States, of whom 30 821 were injured or killed as the result of work accidents. Of these events, 35 per cent were associated with the operation of coupling freight cars or wagons. Prior to that, between 1877 and 1887, 38 per cent of all railworker accidents involved coupling.[23]

Until Congress passed the Safety Appliance Act in 1893, the task of coupling involved guiding a link tube into a coupler pocket and then inserting a pin into a hole on the tube to hold the link in place. To do this, the crew members had to go between moving cars during coupling and were frequently injured and sometimes killed.

In 1873, Eli J. Janney—a dry goods clerk and former Confederate Army officer—patented a knuckle-style coupler that allowed railworkers to couple and uncouple cars without having to go between them to guide the link and set the pin. Though the market was soon flooded with thousands of patented couplers, Janney's design was clearly the best. In 1888, the Master Car Builders Association adopted the Janney coupler as its standard. Congress, satisfied that the device could do the job safely, passed the Safety Appliance Act in 1893, making the Janney coupler a legal requirement throughout the railroad industry.

The impact was staggering. The accident rate fell dramatically during the 10 subsequent years and by 1902 (only two years after the end of the Act's seven-year grace period), coupling accidents represented a mere 4 per cent of all employee accidents. In absolute terms, coupler-related accidents dropped from nearly 11 000 in 1892 to just over 2000 in 1902, even though the number of railroad employees had steadily increased during that decade—from 873 602 in 1893 to 1 189 315 in 1902. Interestingly, there was no reduction in the proportion of railroad employees involved in accidents. This remained steady at around 3–4 per cent of the total workforce in any one year.

Other conspicuously successful safety laws have involved the introduction of speed limits on the roads. The imposition of the 30 mph limit on British roads in the 1930s brought about the largest single reduction in road casualties. The 55 mph nationwide speed limit (previously 70 mph) introduced in the United States in 1974 in response to the oil crisis reduced the fatal crash rate by 34 per cent. The subsequent raising of this speed limit to 65 mph on rural Interstate highways in 1987 led to an increase in crash fatalities of 16 per cent.

In Britain, the wearing of seat belts by drivers and front-seat passengers became a legal requirement on 31 January 1983. A comparison between February–December 1982 and February–December 1983 reveals a 23 per cent drop in fatalities and a 26 per cent reduction in serious injuries. A subsequent study found a fall-off in serious injuries (including fatalities) of 23 per cent for car drivers and of 30 per cent for front-seat passengers—the individuals directly affected by the law.

We could, of course, continue to catalogue the beneficial effects of other safety laws, such as the wearing of crash helmets by motorcyclists—and the adverse effects of repealing these laws in certain states in the USA—but the point has been made. Legislation targeted at restricting particular types of unsafe acts has had a very large impact. But these are all *individual accidents* in which the people at risk, the hazards and the dangerous situations are well known.

Unfortunately, the same degree of specificity—or even understanding—does not exist for *organizational accidents*. So, the question remains: how can regulators function more effectively to limit the occurrence of these catastrophic yet infrequent events? To address this question, we first need to consider the nature of the regulatory process and then look at how key events, such as the UK's Health and Safety at Work Act (1974) and the Cullen Report on *Piper Alpha*, have changed the way in which safety laws are framed and enforced.

Autonomy and Dependence as Constraints on the Regulatory Process

Regulatory authorities are agents of social control. The organizational literature is rich in theoretical accounts of this control process, but the one favoured here is that presented by Diane Vaughan in her detailed analysis of the *Challenger* accident.[24] What makes Vaughan's notions of organizational autonomy and interdependence particularly relevant to our present concerns is that they were developed to elucidate the regulatory problems that can, and did, contribute to an organizational accident.

The essence of her argument is that the regulatory process—discovery, monitoring, investigation and sanctioning—is inevitably constrained by the interorganizational relations existing between the regulatory body and the regulated company. These, in turn, lead to relationships based more upon bargaining and compromise than threats and sanctions. The fact that both the regulator and the regulated are autonomous, existing as separate and independent entities, poses special problems for the regulator. All organizations are, to varying degrees, insulated from the outside world by their physical structure, culture and measures for limiting the leakage of sensitive facts. Information passing outwards gets filtered. Moreover, organizations tend to be highly selective in their transactions with external organizations, and especially with regulators.

Regulators, for their part, attempt to penetrate the boundaries of the regulated organizations by requesting certain kinds of information and by making periodic site visits. But these strategies can only provide isolated glimpses of the organization's activities. Size, complexity, the peculiarities of organizational jargon (as at NASA, for example), the rapid development of technology and, on occasions, deliberate obfuscation all combine to make it difficult for the regulator to gain a comprehensive and in-depth view of the way in which an organization really conducts its business. And, being themselves members of an autonomous organization with its own agenda, individual regulators confronting these difficulties are likely to give their immediate supervisors the impression that they know more about the regulated organization than is actually the case. To confess that they cannot penetrate to the heart of their assigned organization's dealings is to admit that they lack the necessary investigative skills, or are not doing the job diligently enough, or both. Regulators, too, have careers.

In an effort to work around these obstacles, regulators tend to become dependent upon the regulated organizations to help them acquire and interpret information. Such interdependence can undermine the regulatory process in various ways. The regulator's knowledge of the nature and severity of a safety problem can be manipulated by what the regulated organization chooses to communicate and how this material is presented. Regulators, being human beings, tend to establish personal relationships with the regulated— they get to like the people they oversee and come to sympathize with their problems on a personal level—and this sometimes compromises their ability to identify, report or sanction violations.

Bad relations consume limited resources, take up valuable time and are unpleasant and often counterproductive—particularly when the internal sources of information dry up. As a result, both the regulator and the regulated generally try to avoid adversarial encounters, favouring negotiation and bargaining over conflict and confrontation. Diane Vaughan summarizes the situation as follows:

> Situations of mutual dependence are likely to engender continual negotiation and bargaining rather than adversarial tactics, as each party tries to control the other's use of resources and conserve its own. To interpret the consequences of this negotiation and bargaining (e.g., the

'slap-on-the-wrist' sanction or no sanction at all) as regulatory 'failures' is to miss the point. Compromise is an enforcement pattern systematically generated by the structure of inter-organizational regulatory relations.[25]

The Move Towards Self-regulation

The past two or three decades have seen a marked change in the way safety legislation is framed in many industrialized countries. Putting it very simply, there has been a shift away from laws that specify the means by which safe working should be achieved to laws that focus on the attainment of certain safety goals. Instead of rules that prescribe the precise steps to be taken by individuals or organizations, leaving little or no discretion for deviation, the current trend is towards rules that emphasize the required outcomes of safety management, allowing considerable freedom on the part of the operators of hazardous technologies to identify the means by which these ends will be achieved.

Achieving safety goals assumes two basic organizational functions.[26] First, it requires reducing the probability of accident occurrence by identifying and eliminating the conditions that cause accidents. Second, it involves reducing the harmful consequences caused by the accidents that still occur—by establishing accident mitigation and emergency response procedures, by reducing those causal factors likely to increase the injury and damage and by moving potential victims and assets out of the exposure zones. Both of these functions are strongly influenced by what the technology or the society regards as safe enough. There are a number of different approaches:

- *The feasibility or ALARP approach* (risks to be kept as low as reasonably practicable)—safety exists at the point beyond which it is neither technologically nor commercially feasible for the organization to do more.
- *The comparative risk approach*—safety is determined on the basis of comparisons to other risks that the society voluntarily accepts (for example, choice of air travel, road transport, smoking and so on).
- *The de minimis approach*—safety exists when the risks are regarded as trivial, commonly taken as 10^{-6} or better.
- *The zero-risk approach*—safety exists only when there is no risk of an accident with harmful consequences.

As the Boston law professor, Michael Baram, has pointed out:

Irrespective of the concept invoked to define what safety is at a particular point in time, as a society progresses, it demands a higher degree of safety. Thus, safety is a target moving continuously towards zero risk, except for interruptions during times of economic distress or high unemployment.[27]

In the brief account of the move towards self-regulation given below, we will focus on the British scene, though comparable developments have occurred elsewhere—see, for example, the Emergency Planning and Community Right to Know Act, passed by the US Congress in 1986, and the Clean Air Act Amendments, enacted by Congress in 1990.

Throughout the first half of this century, safety legislation in Britain was firmly rooted in the 'Factories Act Model'.[28] This was characterized by its fragmentation and its lack of workforce involvement. Since the mid-nineteenth century, the tendency had been for the passage of highly focused Acts of Parliament involving the detailed prescriptive regulation of specific work activities. The lack of workforce involvement was a legacy of the Factories Act tradition that imposed duties on employers, while workers' participation in the safety effort was seen mainly as a matter of discipline to be enforced by the employers.

By the late 1960s this unwieldy legislative edifice, with one set of industry-related rules piled upon another, was beginning to crumble. A rapidly rising rate of industrial accidents, together with sweeping changes in technology and materials, created a widespread demand for new health and safety legislation. The government established a Commission of Inquiry under Lord Robens, hitherto the Chairman of the National Coal Board. The recommendations of the Robens Report later became enacted—with remarkably little governmental tinkering—as the Health and Safety at Work Act (HSW Act) in 1974. It was this Act that initiated the move towards self-regulation.

The Robens Committee identified a number of major defects in the existing statutory system for promoting health and safety at work. The first, and the most fundamental, of these was that there was too much law. This had the effect of 'conditioning people to think of health and safety at work as in the first and most important instance a matter of detailed rules imposed by external agencies'.[29] The second defect was that much of the existing law was unsatisfactory. It was unintelligible, obsolete in parts and largely concerned with physical circumstances rather than with 'the attitudes, capacities and performance of people and the efficiency of the organizational systems within which they work'.[30] The third problem was the fragmentation of administrative jurisdictions and the absence of a comprehensive official provision for health and safety at work. The Report stated:

> We need a more effectively self-regulating system ... It calls for better systems of safety organization, for more management initiatives, and for more involvement of work people themselves. The objectives of future policy must therefore include not only increasing the effectiveness of the state's contribution to safety and health at work but also, and more importantly, creating conditions for more effective self-regulation.[31]

The second chapter of the Robens Report also made the following pertinent observation:

> Promotion of health and safety at work is an essential function of good management. ... Good intentions at the board level are useless if managers further down the chain and closer to what happens on the shop floor remain preoccupied exclusively with production problems.[32]

The HSW Act set in motion the progressive replacement of the existing health and safety legislation by a system of regulations and approved codes of practice. It imposed on an employer a duty 'to ensure, so far as is reasonably practicable, the health, safety and welfare at work of all his employees' and 'to conduct his undertaking in such a way as to ensure, so far as is reasonably practicable, that persons not in his employment who may be affected thereby are not thereby exposed to risks to their health and safety.' (ss. 2 and 3). It created two new bodies: the Health and Safety Commission (HSC), with responsibility for achieving the general purposes of the Act, and the Health and Safety Executive (HSE), charged with the enforcement of health and safety policy. Excluded from the province of the HSE at that time were offshore operations, consumer and food safety and transport other than that of hazardous goods. In 1976 the Act was extended to cover workers in the offshore oil and gas industry, but the responsibility for regulation was placed with the Department of Energy (that came into existence in 1974) in view of their specialist knowledge and experience of this relatively new area of operations.

To summarize: Unlike the Factories Act, the HSW Act did not go into any great detail with regard to particular accident-producers (for example, machinery, hoists and lifts, gas holders and the like). Instead, it provided broad guidelines for the duties of employers, employees, suppliers and users. It also established the processes for creating future regulations and codes of practice and set out the frameworks for a policy-making body (the HSC) and a regulatory body (the HSE).

Between 1971 and 1980 there was a clear downward trend in overall numbers for both fatal and non-fatal accidents. The trend was particularly marked after 1973 and from 1978 onwards when the picture was one of more or less continuous decline in the number of industrial accidents. It should be noted, however, that this was also part of a long-term decline in industrial accidents that had been in progress since the turn of the century. But from 1980–81 onwards, the UK industrial accident rate levelled out and then began to increase. More people were being seriously injured in the manufacturing and construction industries at the end of the 1981–85 period than in the beginning. The causes of this upturn are still obscure.

In 1984 the HSE issued the Control of Industrial Major Hazards (CIMAH) Regulations covering all onshore major hazard installations. This required that the operator should provide the HSE with a written report on the safety of the installation, commonly known as the Safety Case. These regulations had their origins in the Flixborough disaster in 1974 and were confined to the European Commission's (EC's) Directive on Major Accident Hazards, commonly termed the Seveso Directive. Their requirements included:

- The demonstration of safe operation.
- The notification of major accidents.
- A written report (the Safety Case).
- Updating of the report following any significant changes.
- An obligation to supply the HSE with further information.
- Preparation of an on-site emergency plan.
- Provision of relevant information to the public.
- An emergency plan produced by the local authority.

On receipt of a Safety Case, the HSE assembled a team of varied specialists to establish that all the required information had been provided and to identify any matters of immediate concern. Any such matters were then pursued by letter or by site visits. Once satisfied that the Safety Case complied with the requirements, the HSE then used it as a basis for their subsequent inspection strategy. It should be emphasized that this was not a licensing or approval process that could be seen as transferring some of the responsibility for health and safety to the HSE. This responsibility lay squarely with the operator, and considerable importance was attached by the HSE to management and organizational issues.

One of the most important sources of UK law since the HSW Act has been the Control of Substances Hazardous to Health (COSHH) Regulations, implemented in 1988 in response to an EC Directive on Hazardous Agents. This sets out the guidelines for measures necessary to protect the health not only of employees but of others. The

COSHH Regulations affected virtually every workplace. They also adopted a strategy that was in marked contrast to the HSW Act—namely that, among other things, employers were required to be aware of the properties of over 5000 substances and to identify those which have relevance to their particular workplaces. They also laid down a 'rolling procedure' for the identification, assessment and control of risk. This entailed the following steps:

- Making a list of all hazardous substances, identifying points of use and carrying out risk assessments—using outside experts, if necessary.
- Controlling exposure to substances by means of ventilation, improved facilities for washing, and using less hazardous substances where possible.
- Testing of respiratory equipment, ventilation and the like, and keeping careful records of these examinations.
- Monitoring exposure through the use of valid occupational hygiene techniques.
- Carrying out health surveillance of employees exposed to risk.
- Training employees in the nature of the risks and the precautions to be taken.

The next major development followed directly from the recommendations set out in the second volume of the Cullen Report on the *Piper Alpha* disaster, published in November 1990. Building on the recent legislation, Lord Cullen proposed a new regulatory mode for offshore operations. He described this as follows:

> I am satisfied that operators of installations ... should be required to carry out a formal safety assessment of major hazards, the purpose of which would be to demonstrate that the potential major hazards of the installation and the risks to the personnel thereon have been identified and appropriate controls provided. This is to assure the operators that their operations are safe. However it is also a legitimate expectation of the workforce and the public that operators should be required to demonstrate this to the regulatory body. The presentation of the formal safety assessment should take the form of a Safety Case, which would be updated at regular intervals and on the occurrence of a major change of circumstances. [33]

The Cullen Report recommended that the regulation of offshore safety should be carried out by 'a discrete division of the HSE which is exclusively devoted to offshore safety'. It also proposed that the HSE should employ a specialist inspectorate that had 'a clear identity and strong influence in the HSE', and that it should be headed by a chief executive who reported directly to the Director General. The chief executive should also be a member of the HSE's senior management board.

This model of Formal Safety Assessments (FSAs) presented as Safety Cases subsequently spread to other industries. For example, prompted in part by the *Estonia* disaster in 1994, the Marine Safety Agency (MSA), in September 1995, created a well funded research programme to investigate the formal safety assessment of shipping. An FSA was seen by the MSA as comprising five steps:[34]

- Identification of hazards.
- Assessment of risks associated with those hazards.
- Consideration of alternative ways of managing those risks.
- Cost–benefit assessment of alternative risk management options.
- Decisions on which options to select.

One further development is worth a brief mention—the upsurge within Britain of an 'ideology of deregulation'. In 1992 the Prime Minister made a speech urging UK regulators to take a more relaxed approach to enforcing the 'excessive detail' of EU regulations. Three months later, the Secretary of State for Trade and Industry set up two inquiries into 'red tape'. One was to look into opportunities for lessening the 'burden' of EU regulations, and the second was tasked with reviewing, simplifying and, where possible, abolishing some 7000 regulations on business. Within two months, the HSE announced its plans to review over 400 health and safety regulations with the intention of lessening their burden upon small businesses. Meanwhile, Parliament passed the Contracting Out and Deregulation Bill, granting the Secretary of State for Employment new powers to repeal safety legislation.

Running in parallel with these moves towards deregulation, there has been—predictably—a steady erosion of the HSC/HSE's budget. This amounted to reductions of 2.6 per cent and 5 per cent in successive years, a fall of some £15 million. Equally predictably, an all-party Parliamentary committee has recently criticized the HSE for conducting too few inspections and, most particularly, for failing to meet its targets with regard to examining outstanding Safety Cases called for under the CIMAH regulations relating to hazardous installations.

Although by no means a definitive review, enough has probably been said to cover the main steps in the move towards self-regulation—and deregulation—in the United Kingdom. It is now time to consider the implications of these changes for the regulatory authorities. Will these measures be effective in limiting the occurrence of organizational accidents? Has it made the regulator's job any easier?

The Pluses and Minuses of the Move to Self-regulation

Any measure that shifts the onus for maintaining safe work practices on to the organizations directly concerned has to represent an enormous 'plus' in the struggle to limit the occurrence of organizational accidents. So also is the related demand upon them to take a close, continuing and proactive interest in all the varied factors affecting the safety of their installations. Most technological operations, even very complex ones, are relatively simple in comparison to the task of maintaining safe working conditions. As noted in Chapter 4, there are not enough trees in the rain forests to carry all the procedures necessary to guarantee safe operations. Safety, as we have seen, is a 'dynamic non-event' that depends crucially upon a clear understanding of the interactions between many different underlying processes. The long-term safety benefits of being forced to grapple with these enormously difficult—and still unresolved—sociotechnical issues are undoubtedly greater than any number of purely technical 'fixes'. The process is more valuable than the product. And, as the Cullen Report notes, many operators have found the exercise of producing a Safety Case valuable: 'Often it would be the first time that a report had been made of the major hazard aspects of the organization. Many stated that the exercise had led them to make changes in their approach and improvements to systems and procedures.'[35]

While there can be little doubt that this legislative drive towards self-regulation has definitely got its heart in the right place when it comes to reducing the likelihood of organizational accidents, its benefits for the regulatory bodies are less certain. Let us examine some of the problems.

Traditionally, regulators worked to ensure compliance with safety rules laid down by some legislative authority. No matter how fragmented, obsolescent or externalized these rules were, they nonetheless represented an agreed standard—at least at that time—against which to determine whether or not a particular work practice or hazardous installation was in violation of this or that regulation. In the new self-determining climate, regulators are still required to look out for deviations, but of quite a different kind. Now they have to inspect for departures from a Safety Case that is expressed in far more general terms, that can vary widely from organization to organization (according to their means) and for which they must take some direct responsibility—since it would not exist as a frame of reference had the regulatory authority not approved it in the first place. As we have seen, spotting, monitoring and sanctioning violations were difficult enough in the past, but now the responsibility placed upon the regulator is very great indeed. Not only do they have to police compliance with a variety of Safety Cases, they also need a very clear idea of what constitutes an adequate Safety Case, and this is by no means an easy task in our present limited state of knowledge.

In order to judge the adequacy of a Safety Case in something other than a cursory checklist fashion, regulators are now required to have a comprehensive—one could almost say god-like—appreciation of all the many factors contributing to both individual and organizational accidents. While the physical origins of the former were largely enshrined in earlier legislation, the various ways in which human, technical and organizational factors can combine to produce the latter are still not fully known and are perhaps ultimately unknowable—since each major organizational accident seems to throw up a fresh set of surprises. This additional evaluative burden is not lightened by the fact that most regulatory staff possess expertise in technical and operational matters rather than in human and organizational factors. While the 80:20 rule (80 human/organizational to 20 technical) may apply to accident causation, it certainly does not reflect the balance of expertise in most regulatory authorities. Even the largest of them have only a sprinkling of individuals with formal training in these 'softer' disciplines—and probably rightly so, since they have yet to deliver the goods with regard to agreed theory and practice. So, even if the causally appropriate 80:20 expertise ratio were achieved, it would not necessarily solve the problem.

The situation for the regulatory authority would become even more difficult should one of its overseen organizations suffer a major accident. The subsequent investigation could turn up one of two things: either that the organization's performance was in compliance with its Safety Case, or that the accident was due in part to violations of the Safety Case. The former could be judged as stemming from shortcomings in the regulator's evaluation process—the Safety Case should not have been approved in the first place—while the latter is likely to be viewed as a failure of regulatory surveillance. Damned if they do and damned if they don't!

A Possible Model for the Regulatory Process

Regulators are uniquely placed to function as one of the most effective defences against organizational accidents. They are located close to the boundaries of the regulated system, but they are not of it. This grants them the perspective to identify unsatisfactory practices and poor equipment that the organization has grown accustomed to or works around. Regulators are specialists in the technology in question. Indeed, many of them have been recruited from the regulated industries. Regulators have been trained and are highly experienced in identifying technical inadequacies and formal systemic weaknesses.

And they possess the investigative access and sanctions necessary to enforce their decisions.

What regulators often lack, however, are the tools, training and the resources to confront the all-important human and organizational factors and to monitor the insidious accumulation of the latent conditions that can subsequently combine to penetrate the system's defences. The same also applies to the policy-makers. The past decade has seen isolated flurries of post-disaster legislation, when what was needed were the statutory and regulatory structures necessary to forestall future accidents rather than trying to prevent the previous ones.

This penultimate section draws heavily on the principles and techniques presented in Chapters 6 and 7 in order to sketch out a model for a regulatory process that, it is believed, could be effective in limiting the occurrence of organizational accidents. The regulatory model outlined below is shaped by three concerns:

- How can regulators deploy their limited resources in the most cost-effective and targeted manner?
- How can regulators bring about the organizational reforms necessary to achieve, and then sustain, optimum levels of 'safety health' on the part of the complex well defended organizations that they oversee?
- Since the absolute criteria for safe operation are rarely known in advance, how can we design a regulatory process that will enable the policy-maker, the regulator and the regulated all to be integral parts of an effective learning cycle?

Figure 8.1 summarizes the basic elements of the regulatory process as they might appear to front-line inspectors. Site visits generate various indicators, such as instances of non-compliance or deviations from safe working practices. These are the raw data of the regulatory process. They could include, for example, badly maintained or unsuitable items of equipment, lack of briefing or drills, a poor permit-to-work system or an unsatisfactory shift handover arrangement. Each instance generates two kinds of action item: those on the regulated organization to put them right (the dark octagons), and those on the regulator to monitor their rectification and, if this is not done satisfactorily, to impose some sanction (the white octagons).

Figure 8.2 shows the next stage of model development. Here, the basic elements of regulation have been extended to identify the upstream organizational and managerial (O & M) determinants of the local instances. The precise nature of these O & M factors will vary from industry to industry, but the assessed factors will always be less

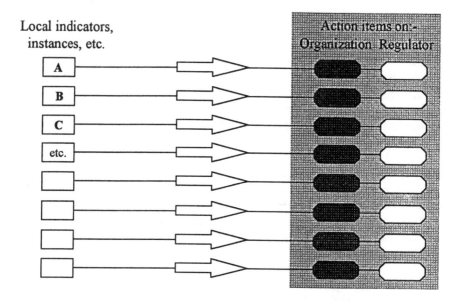

Figure 8.1   The basic elements of the regulatory process

than the total number of possible factors (as discussed in Chapter 7). In this case, only seven O & M factors (represented by the dark rectangles) are considered in the analysis. If, for example, the regulated organization were a small airline, the O & M factors could include crew factors, operational management, maintenance management, safety management, organizational structure, commercial and operational pressures.

It is presumed that each instance is the product of different contributions from each of the O & M factors. The circular nodes in the lower half of the figure are points at which the regulator is required to make an assessment of the impact of each O & M factor on that particular instance. The regulator rates the relative contributions of each O & M factor to each instance on a five-point scale, where 1 = very little influence and 5 = very considerable influence. The associated white rectangles are action items on the regulator to monitor and, if necessary, to assist with any reforms.

These ratings are then summed over all the instances in that particular collection (either from a single inspection or several) to generate an Organizational Factor Profile (see Figure 8.3). The purpose of the profile is to indicate—to both the regulator and the regulated—which of the various upstream factors is most in need of urgent reform. As discussed in Chapter 7, we cannot expect any organization to deal

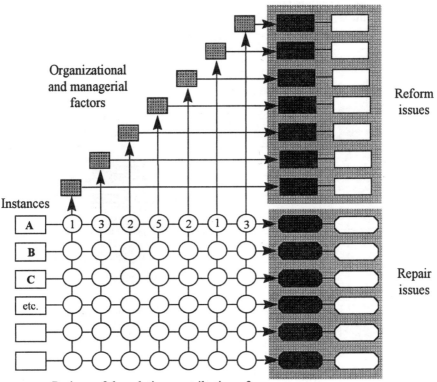

Organizational
and managerial
factors

Reform
issues

Instances

| A | 1 | 3 | 2 | 5 | 2 | 1 | 3 |

| B |

| C |

| etc. |

Repair
issues

Ratings of the relative contribution of
each organizational factor to each instance

**Figure 8.2   Addition of the organizational and managerial (O &
M) factors to the basic elements of the regulatory
process**
Whereas individual instances generate local repair
issues, the action items (on both the regulated and the
regulator) generated by the O & M factors require more
fundamental systemic reforms. The dark rectangles in
the upper half represent action items on the regulated
organization to reform a specific O & M factor. The associated white rectangles are action items on the regulator
to monitor and, if necessary, to assist with these reforms

effectively with all of its problems at once. Later inspections will
yield further instances that, in turn, will generate more organizational factor profiles. A succession of these profiles will allow both
the regulator and the regulated to track the progress of remedial
efforts.

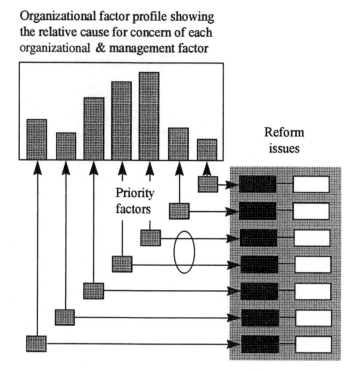

Organizational factor profile showing
the relative cause for concern of each
organizational & management factor

Figure 8.3   An Organizational Factor Profile generated from the ratings summed over all the instances
The purpose of the profile is to identify those two or three factors most in need of reform by the organization (and subsequent close monitoring by the regulator).

Finally, Figure 8.4 attempts to show how the regulatory process modelled here could form part of a wider learning cycle, involving legislators, regulators, and hazardous technologies. The systemic improvements generated by these focused organizational reforms will, it is hoped, come to represent new standards of health and safety at work. This could be incorporated into new legislation which, in turn, would change the regulator's inspection and surveillance criteria—and so on. It will be seen that the local repairs required of regulated organizations become relegated to a subordinate loop. The main emphasis throughout is upon the continuing reform of the source factors influencing a system's intrinsic resistance to its operational hazards.

The philosophy underlying this model is identical to that described for Tripod-Delta (see Chapter 7). Local instances, such as unsafe acts, are merely symptoms of the underlying organizational and manage-

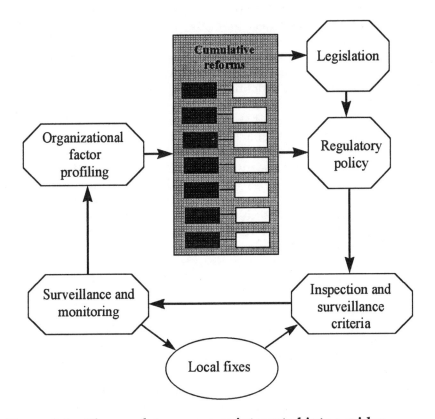

Figure 8.4   The regulatory process integrated into a wider learning cycle

rial pathology. Removing the instances one-by-one will not improve the organization's safety health. That can only be achieved by curing the source problems.

As noted in Chapter 6, maintaining safety, like life, is 'one damn thing after another'. It is to be expected that while each targeted O & M factor is being reformed others will be deteriorating. Nevertheless, the model provides a principled way of tracking these changes and of addressing the latest 'big problems' as each is revealed by the prevailing pattern of local indicators.

The Regulator Deserves a Better Deal

Judging by the failures presented in the first part of the chapter, regulatory authorities are not always able to prevent organizational accidents. But a closer examination shows that they, like other people in the front-line of the safety war, are more often victims than villains. Yes, they make mistakes. But why should they be any different from the rest of the human race? And, like other people, they do not make their errors in isolation. As this chapter has tried to show, the majority of regulatory shortcomings have their origins far upstream from the individual inspector.

Given the current trend of searching for increasingly more remote contributions to organizational accidents, it is inevitable that the regulator's alleged deficiencies should be judged by those with 20:20 hindsight as making significant contributions to a major disaster. The regulators' position vis à vis the affected organization means that they are likely to attract blame from all directions. Standing as they do on the organizational borders of all hazardous technologies, their sphere of responsibility is bound to be implicated in a wide variety of contributing factors. However, if regulators are to be other than convenient scapegoats, they will have to be provided with the legislation, the resources and the tools to do their jobs effectively. They are potentially one of the most important defences against organizational accidents

Societies, just like the operators of hazardous systems, put production before protection. As we have seen, safety legislation is enacted in the aftermath of disasters, not before them. There is little or no political kudos to be gained from bringing about a non-event, although, in the long run, meeting this challenge successfully is likely to be much more rewarding. Every society gets the disasters it deserves. Let's hope that, in the next millennium, the regulators are seen to deserve something better than has so far been the case. Then, perhaps, we will all be safer.

Notes

1   *Report of the Presidential Commission on the Space Shuttle 'Challenger' Accident*, (Washington, DC: Government Printing Agency, 1986). See also the excellent account by H.S.F. Cooper, 'A letter from the Space Center', *The New Yorker*, 10 November 1987.

2   D. Vaughan, *The Challenger Launch Decision*, (Chicago: Chicago University Press, 1996). Malcolm Gladwell's article 'Blowup', *The New Yorker*, 22 January 1996 reviews this book and places it in its broader theoretical context. Her most recent publication on this topic is 'The trickle-down effect: policy decisions, risky work, and the *Challenger* tragedy', *California Management Review*, 39, 1997, pp. 8–102.

3   D. Vaughan, 'Autonomy, interdependence, and social control: NASA and the Space Shuttle *Challenger*', *Administrative Science Quarterly*, 35, 1990, pp. 225–33.

4   D. Fennell, *Investigation into the King's Cross Underground Fire*, (Department of Transport, London: HMSO, 1988).

5   Ibid., pp. 146--7.
6   The Hon. Lord Cullen, *Public Inquiry into the Piper Alpha Disaster*, (Department of Energy, London: HMSO, 1990).
7   Ibid., vol. 1, pp. 239–41.
8   Ibid., vol. 1, p. 253.
9   Mr Justice V.P. Moshansky, *Commission of Inquiry into the Air Ontario Crash at Dryden, Ontario. Final Report*, Vol. 1, (Ottawa: Ministry of Supply and Services, 1992), pp. 5–6. See also the analysis by Captain Dan Maurino in D. Maurino *et al.*, *Beyond Aviation Human Factors*, (Aldershot: Avebury), 1995, pp. 57–85.
10 Mr Justice V.P. Moshansky, op. cit., vol. 3., p. 914.
11 Ibid., vol. 1, p. xxviii.
12 Ibid., vol. 1, p. 7.
13 To avoid confusion with other civil aviation authorities, the Australian CAA will be referred to as the ACAA in the text.
14 Bureau of Air Safety Investigation (BASI), *Piper PA31-350 Chieftain Young NSW 11 June 1993*, Investigation Report 9301743, (Canberra, ACT: Bureau of Air Safety Investigation, 1994).
15 Editorial, *The Sydney Morning Herald*, 7 October 1994.
16 M. Riley, 'Police to investigate CAA link to Seaview', *The Sydney Morning Herald*, 14 October 1994.
17 M. Taylor, 'Brereton strips CAA safety role', *The Canberra Times*, 13 October 1994.
18 *Financial Review*, 11 October 1994.
19 A. Bryant, 'Marginal safety: a special report', *The New York Times*, 13 November 1995.
20 E. Pooley, 'Nuclear warriors', *Time*, 4 March 1996.
21 This can also be termed a 'pond', as in the UK, for example.
22 E. Pooley, op. cit.
23 The information relating to the decline in coupling accidents is taken from *The United States Law Week*, 27 February 1996, **64**, p. 4112. I am most grateful to Michael Baram for sending it to me, although I am not sure that he would entirely approve of the use to which I have put it.
24 D. Vaughan, 'Autonomy, interdependence, and social control: NASA and the space shuttle *Challenger*', *Administrative Science Quarterly*, **35**, 1990, pp. 225–33.
25 Ibid., p. 227.
26 M. Baram, 'The use of rules to achieve safety: introductory remarks', paper presented to the Workshop on the Use of Rules to Achieve Safety, Bad Homburg, Germany, 6 May 1993.
27 Ibid., p. 3.
28 S. Dawson, P. Willman, M. Bamford and A. Clinton, *Safety at Work: The Limits of Self-Regulation*, (Cambridge: Cambridge University Press, 1988).
29 The Hon. Lord Cullen, op. cit., p. 256.
30 The Hon. Lord Cullen, op. cit., p. 257.
31 The Hon. Lord Cullen, op. cit., p. 257.
32 Lord Robens, Safety and Health at Work, *Report of the Committee 1970–72*, Vol 1, London: HMSO, 1972, p. 41.
33 The Hon. Lord Cullen, op. cit., vol. 1, p. 3.
34 Marine Safety Agency (MSA), *Formal Safety Assessment of Shipping: Programme Overview*, (London: Marine Safety Agency, 15 September, 1995).
35 The Hon. Lord Cullen, op. cit., p. 277.

9 Engineering a Safety Culture

The Scope of the Chapter

Few phrases occur more frequently in discussions about hazardous technologies than *safety culture*. Few things are so sought after and yet so little understood. However, should it be thought that the current preoccupation with safety culture is just another passing fad, consider the following facts. Commercial aviation is an industry that possesses an unusual degree of uniformity worldwide. Airlines across the globe fly much the same types of aircraft in comparable conditions. Flight crews, air traffic controllers and maintenance engineers are trained and licensed to very similar standards. Yet, in 1995, the risks to passengers (the probability of becoming involved in an accident with at least one fatality) varied by a factor of 42 across the world's air carriers, from a 1 in 260 000 chance of death or injury in the worst cases to a 1 in 11 000 000 probability in the best cases.[1] While factors such as national and company resources will play their part, there can be little doubt that differences in safety culture are likely to contribute the lion's share to this enormous variation.

We first encountered the term 'safety culture' in Chapter 2 when making the distinction between pathological, bureaucratic and generative organizations. It cropped up again in Chapter 6 in regard to the motive forces that drive an organization towards a state of maximum resistance to its operational hazards. The present chapter focuses mainly on three questions: What is an organizational culture? What are the main ingredients of a safety culture? And, most importantly, how can it be engineered? The term 'engineered' is deliberate. But it is not meant in the traditional sense of developing more sophisticated gadgetry. Rather, we will be discussing the application of *social engineering*.

This book has sought to argue that most of the effective solutions to human performance problems are more the province of the technical manager (and the regulator) than the psychologist since they

191

concern the conditions under which people work rather than the human condition itself. The main message of this chapter is that the same general principle also applies to the acquisition of an effective safety culture (hereafter, the phrase 'safety culture' will be taken to mean an effective safety culture). Whereas national cultures arise largely out of shared values, organizational cultures are shaped mainly by shared practices (a claim that is developed in the next section). And it is these practices that will be the focus of this chapter.

Many people talk as if a safety culture can only be achieved through some awesome transformation, akin to a religious experience. This chapter takes the opposite view, arguing that a safety culture can be socially engineered by identifying and fabricating its essential components and then assembling them into a working whole. It is undoubtedly true that a bad organizational accident can achieve some dramatic conversions to the 'safety faith', but these are all too often shortlived. A safety culture is not something that springs up ready-made from the organizational equivalent of a near-death experience, rather it emerges gradually from the persistent and successful application of practical and down-to-earth measures. There is nothing mystical about it. Acquiring a safety culture is a process of collective learning, like any other. Nor is it a single entity. It is made up of a number of interacting elements, or ways of doing, thinking and managing that have enhanced safety health as their natural byproduct.

What is an Organizational Culture?

To those with a 'hard' engineering background, many attempts to describe the nature of organizational culture must seem to have the definitional precision of a cloud. There is no standard definition, but here is one that captures most of the essentials with the minimum of fuss:

> Shared values (what is important) and beliefs (how things work) that interact with an organization's structures and control systems to produce behavioural norms (the way we do things around here).[2]

Until the 1980s, 'culture' was a term applied more to nationalities than to organizations. 'Organizational culture' became an essential part of 'management speak' largely as the result of two widely read books: *Corporate Culture* (by Terrence Deal and Allan Kennedy)[3] and *In Search of Excellence* (by Thomas Peters and Robert Waterman),[4] both published in 1982.

The latter book introduced the notion of *cultural strength,* and observed 'Without exception, the dominance and coherence of culture proved to be an essential quality of the excellent companies'.[5] Fifteen years later, there are some who might doubt this assertion (particularly those laid off from the 'excellent' companies), but few would argue with the idea that a strong culture is one in which all levels of the organization share the same goals and values. To quote Peters and Waterman again, 'In these [strong culture] companies, people way down the line know what they are supposed to do in most situations because the handful of guiding values is crystal clear'.[6]

Organizational theorists have described a number of negative or dysfunctional cultures. One such 'bad' culture is characterized by what psychologists have termed *learned helplessness,* describing a condition in which people learn that attempts to change their situation are fruitless so that they simply give up trying: 'The energy and will to resolve problems and attain goals drains away.'[7] Another counterproductive organizational strategy is *anxiety-avoidance.* When such an organization discovers a technique for reducing its collective anxiety, it is likely to be repeated over and over again regardless of its actual effectiveness.

> The reason is that the learner will not willingly test the situation to determine whether the cause of the anxiety is still operating. Thus all rituals, patterns of thinking or feeling, and behaviours that may originally have been motivated by a need to avoid a painful, anxiety-provoking situation are going to be repeated, even if the causes of the original pain are no longer acting, because the avoidance of anxiety is, itself, positively reinforcing.[8]

Both learned helplessness and repetitive anxiety-avoidance are likely to assist in driving the blame cycle, described in Chapter 7. People feel helpless in the face of ever-present dangers and, while the familiar reactions to incidents and events, such as 'write another procedure' and 'blame and train,' may not actually make the system more resistant to future organizational accidents, they at least serve the anxiety-reducing function of being seen to do something—and blaming those at the sharp end deflects blame from the organization as a whole.

There is a controversy among social scientists as to whether a culture is something an organization 'has' or whether it is something an organization 'is'. The former view emphasizes management's power to change culture through the introduction of new measures and practices, while the latter sees culture as a global property that emerges out of the values, beliefs and ideologies of the organization's entire membership. The former approach is favoured by managers and management consultants, while the latter is preferred by academics and social scientists. This chapter stands with the managers and agrees with the organizational anthropologist, Geert Hofstede, when he wrote:

> On the basis of [our] research project, we propose that practices are features an organization *has*. Because of the important role of practices in organizational cultures, the ['has' approach] can be considered as somewhat manageable. Changing collective values of adult people in an intended direction is extremely difficult, if not impossible. Values do change, but not according to someone's master plan. Collective practices, however, depend on organizational characteristics like structures and systems, and can be influenced in more less predictable ways by changing these.[9]

Although the idea of a safety culture has existed since 1980, it was given an authoritative boost by the International Atomic Energy Agency when they published a report in 1988, elaborating the concept in detail. They defined safety culture as: '... that assembly of characteristics and attitudes in organizations and individuals which establishes that, as an overriding priority, nuclear plant safety issues receive the attention warranted by their significance'.[10] Unfortunately, this is something of a 'motherhood' statement specifying an ideal but not the means to achieve it. A more useful definition, which is worth quoting in full, has been given by the UK's Health and Safety Commission in 1993:

> The safety culture of an organization is the product of individual and group values, attitudes, competencies, and patterns of behaviour that determine the commitment to, and the style and proficiency of, an organization's health and safety programmes. Organizations with a positive safety culture are characterized by communications founded on mutual trust, by shared perceptions of the importance of safety, and by confidence in the efficacy of preventive measure.[11]

While remaining in sympathy with this definition, this chapter emphasizes the critical importance of an effective safety information system—the principal basis of an *informed culture*. It must be stressed again that our primary concern in this book is not with traditional health and safety measures that are directed, for the most part, at the prevention of *individual* work accidents. Our focus is upon the limitation of *organizational* accidents, and it is this that has shaped the arguments set out below.

The Components of a Safety Culture

The main elements of a safety culture and their various interactions are previewed below. Each subcomponent will be discussed more fully in succeeding sections.

- As indicated in Chapter 6, an ideal safety culture is the engine that continues to propel the system towards the goal of maximum safety health, regardless of the leadership's personality or current commercial concerns. Such an ideal is hard to achieve in the real world, but it is nonetheless a goal worth striving for.
- The power of this engine relies heavily upon a continuing respect for the many entities that can penetrate and breach the defences. In short, its power is derived from not forgetting to be afraid.
- In the absence of bad outcomes, the best way—perhaps the only way—to sustain a state of intelligent and respectful wariness is to gather the right kinds of data. This means creating a safety information system that collects, analyses and disseminates information from incidents and near-misses as well as from regular proactive checks on the system's vital signs (see Chapter 7). All of these activities can be said to make up an *informed culture*—one in which those who manage and operate the system have current knowledge about the human, technical, organizational and environmental factors that determine the safety of the system as a whole. In most important respects, an informed culture *is* a safety culture.
- Any safety information system depends crucially on the willing participation of the workforce, the people in direct contact with the hazards. To achieve this, it is necessary to engineer a *reporting culture*—an organizational climate in which people are prepared to report their errors and near-misses.
- An effective reporting culture depends, in turn, on how the organization handles blame and punishment. A 'no-blame' culture is neither feasible nor desirable. A small proportion of human unsafe acts are egregious (for example, substance abuse, reckless non-compliance, sabotage and so on) and warrant sanctions, severe ones in some cases. A blanket amnesty on all unsafe acts would lack credibility in the eyes of the workforce. More importantly, it would be seen to oppose natural justice. What is needed is a *just culture*, an atmosphere of trust in which people are encouraged, even rewarded, for providing essential safety-related information—but in which they are also clear about where the line must be drawn between acceptable and unacceptable behaviour.

- The evidence shows that high-reliability organizations—domain leaders in health, safety and environmental issues—possess the ability to reconfigure themselves in the face of high-tempo operations or certain kinds of danger. A *flexible culture* takes a number of forms, but in many cases it involves shifting from the conventional hierarchical mode to a flatter professional structure, where control passes to task experts on the spot, and then reverts back to the traditional bureaucratic mode once the emergency has passed. Such adaptability is an essential feature of the crisis-prepared organization and, as before, depends crucially on respect—in this case, respect for the skills, experience and abilities of the workforce and, most particularly, the first-line supervisors. But respect must be earned, and this requires a major training investment on the part of the organization.
- Finally, an organization must possess a *learning culture*—the willingness and the competence to draw the right conclusions from its safety information system, and the will to implement major reforms when their need is indicated.

The preceding bullet points have identified four critical subcomponents of a safety culture: a *reporting culture*, a *just culture*, a *flexible culture* and a *learning culture*. Together they interact to create an *informed culture* which, for our purposes, equates with the term 'safety culture' as it applies to the limitation of organizational accidents.

Engineering a Reporting Culture

On the face of it, persuading people to file critical incident and near-miss reports is not an easy task, particularly when it may entail divulging their own errors. Human reactions to making mistakes take various forms, but frank confession does not usually come high on the list. Even when such personal issues do not arise, potential informants cannot always see the value in making reports, especially if they are sceptical about the likelihood of management acting upon the information. Is it worth the extra work when no good is likely to come of it? Moreover, even when people are persuaded that writing a sufficiently detailed account is justified and that some action will be taken, there remains the overriding problem of trust. Will I get my colleagues into trouble? Will I get into trouble?

There are some powerful disincentives to participating in a reporting scheme: extra work, scepticism, perhaps a natural desire to forget that the incident ever happened, and—above all—lack of trust and, with it, the fear of reprisals. Nonetheless, many highly effective reporting programmes do exist. What can we learn from them? How have they engineered their success?

In what follows, we will briefly look at the 'social engineering' details of two successful aviation reporting programmes, one operating at a national level and the other within a single airline. These are NASA's Aviation Safety Reporting System (ASRS) and the British Airways Safety Information System (BASIS). In this, we will rely heavily on the work of two people—Dr Sheryl Chappell of NASA, and Captain Mike O'Leary of British Airways[12]—each of whom has been closely involved in the design and management of these programmes. Our purpose here is not to consider the programmes themselves in any detail, but to abstract from them the 'best practices' for achieving a reporting culture. Throughout, we will concentrate on the issue of how valid reporting may be promoted. Although we are dealing exclusively with aviation reporting schemes, the basic 'engineering' principles can be applied in any domain. Indeed, reporting programmes in other domains—particularly in medicine—are partly modelled on pioneering aviation schemes such as critical incident reporting.

Examination of these successful programmes indicates that five factors are important in determining both the quantity and the quality of incident reports. Some are essential in creating a climate of trust, others are needed to motivate people to file reports. The factors are:

- Indemnity against disciplinary proceedings—as far as it is practicable.
- Confidentiality or de-identification.
- The separation of the agency or department collecting and analysing the reports from those bodies with the authority to institute disciplinary proceedings and impose sanctions.
- Rapid, useful, accessible and intelligible feedback to the reporting community.
- Ease of making the report.

The first three items are designed to foster a feeling of trust. O'Leary and Chappell explain the need:

> For any incident reporting programme to be effective in uncovering the failures which contribute to an incident, it is paramount to earn the trust of the reporters. This is even more important when there is a candid disclosure of the reporter's own errors. Without such trust, the report will be selective and will probably gloss over pivotal human factors information. In the worst case—that in which potential reporters have no trust in the safety organization—there may be no report at

all. Trust may not come quickly. Individuals may be hesitant to report until the reporting system has proved that it is sensitive to reporters' concerns. Trust is the most important foundation of a successful reporting programme, and it must be actively protected, even after many years of successful operation. A single case of a reporter being disciplined as the result of a report could undermine trust and stop the flow of useful reports.[13]

The rationale for any reporting system—and a recurrent theme throughout this book—is that valid feedback on the local and organizational factors promoting errors and incidents is far more important than assigning blame to individuals. To this end, it is essential to protect informants and their colleagues as far as possible from disciplinary actions taken on the basis of their reports. But there will be limits upon this indemnity. These limits are defined most clearly by the Waiver of Disciplinary Action issued in relation to NASA's Aviation Safety Reporting System. Below is an excerpt from the FAA Advisory Circular (AC No. 00-46C) describing how the immunity concept applies to pilots making incident reports.

The filing of a report with NASA concerning an incident or occurrence involving a violation of the Act of the Federal Aviation Regulations is considered by the FAA to be indicative of a constructive attitude. Such an attitude will tend to prevent future violations. Accordingly, although a finding of a violation may be made, neither a civil penalty nor a certificate suspension will be imposed if:

- The violation was inadvertent and not deliberate;
- The violation did not involve a criminal offence, or accident or ... a lack of qualification or competency;
- The person has not been found in any prior FAA enforcement action to have committed a violation of the Federal Aviation Act, or of any regulation promulgated under the Act for a period of 5 years prior to the date of the occurrence; and
- The person proves that, within 10 days after the violation, he or she completed and delivered or mailed a written report of the incident or occurrence to NASA under ASRS.[14]

This formula appears to work. The ASRS reporting rate was high, even at the outset. In the beginning, it averaged approximately 400 reports per month. It now runs at around 650 reports per week and more than 2000 reports per month. In 1995 ASRS received over 30 000 reports.

BASIS has been extended over the years to cover a wide variety of reporting schemes. All flight crew are required to report safety-related events using Air Safety Reports (ASRs). ASRs are not anonymous. To encourage the filing of ASRs, British Airways Flight Crew Order, No. 608 states:

It is not normally the policy of British Airways to institute disciplinary proceedings in response to the reporting of any incident affecting air safety. Only in rare circumstances where an employee has taken action or risks which, in the Company's opinion, no reasonably prudent employee with his/her training and experience would have taken, will British Airways consider initiating such disciplinary action. [15]

Again, the formula seems to work. Its success is suggested by two statistics. First, the ASR filing rate more than trebled between its inception in 1990 and 1995. Second, the combined number of reports assigned to the severe and high risk categories has decreased by two-thirds between the first six months of 1993 and the first half year of 1995.[16]

Another important component of BASIS is the British Airways Confidential Human Factors Reporting Programme, instituted in 1992. While the ASRs provided good technical and procedural information, a need was felt for an information channel that was more sensitive to human factors issues. Each pilot filing an ASR is now invited to complete a confidential human factors questionnaire relating to the incident. The return of the questionnaire is voluntary. The following assurance was given by the senior manager in charge of BA's Safety Services on the front page of the initial version:

I give an absolute assurance that any information you provide will be treated confidentially by Safety Services and that this questionnaire will be destroyed immediately after the data is processed. This programme is accessible only to my Unit.

In its first year of operation, the human factors reporting programme received 550 usable responses.[17] The issues raised in the reports are communicated to management on a regular basis, but great care is taken to separate the important safety issues from the incidents in order to preserve the anonymity of the reports.

Another important input to BASIS comes from the Special Event Search and Master Analysis (SESMA). This by-passes the need for human reporting by monitoring directly the flight data recorders (FDRs) of BA's various aircraft fleets, while at the same time guaranteeing the flight crews complete anonymity. The FDR for each flight is scanned for events that are considered to lie outside safe norms. All events are stored in a BASIS database and the more serious are discussed at a monthly meeting of technical managers and the pilots' union representatives. If the incident is considered to be sufficiently serious, the union representative is required to discuss the matter with the flight crew involved—while still withholding their identities from the management.

When a report is received by NASA's ASRS staff it is processed in the following manner, with great care being taken to preserve the anonymity of the reporter.[18]

- An initial analysis screens out reports involving accidents, criminal behaviour, or those classified as 'no safety content'.
- The report is coded and the reporter de-identified. At this stage, the reporter is also contacted by telephone to confirm receipt and de-identification.
- After a quality check, the information is entered into the ASRS database and the original report destroyed.

The most obvious way of ensuring confidentiality is to have the reports filed anonymously. But, as O'Leary and Chappell point out, this is not always possible or even desirable.[19] The main problems with total anonymity are as follows:

- Analysts cannot contact the informant to resolve questions.
- It is more likely that some managers will dismiss anonymous reports as the work of disaffected troublemakers.
- In small companies, it is almost impossible to guarantee anonymity.

O'Leary and Chappell conclude that removing identities from reports at a later stage—as described above for ASRS—is probably the most workable means of maintaining confidentiality. At a national level, complete de-identification means removing not only the people's names, but also the date, the time, the flight number and the airline name. The criteria for de-identification must be known and understood by all potential reporters.

Another important measure for engendering trust is to separate the organization receiving the reports from both the regulatory body and from the employing company. As in the case of ASRS, the system analysts should ideally have no legal or operational authority over the potential reporters. Reporting systems run by disinterested third parties—such as universities—can also help to earn the trust of reporters. If, like BASIS, the reporting system is internal to a company, the receiving department should be perceived as being completely independent of operational management, thus giving the necessary assurance of confidentiality.

Apart from a lack (or loss) of trust, few things will stifle incident reporting more than the perceived absence of any useful outcome. Both the ASRS and BASIS place great emphasis on the rapid feedback of meaningful information to their respective communities. If an ASRS report describes a continuing hazardous situation—for example, a defective navigation aid, a confusing procedure, or an incorrect chart—an alerting message is sent out immediately to the appropriate authorities so that they can investigate the problem and take the necessary remedial action. (As mentioned earlier, ASRS has no legal or operational authority of its own.) Some 1700 alert bulletins and notices have been issued by the ASRS team since the programme began in 1976. In 1994 there was a 65 per cent response rate to alert bulletins and FYI notices.

The information assembled in the ASRS database is made available to the aviation and research communities in a variety of ways. First, targeted searches can be carried out at the request of companies, agencies or researchers. The information is also disseminated via a newsletter, *Callback*, whose extended readership is estimated at over 100 000, and other ASRS reports. Such newsletters describe safety issues and highlight improvements that have been made as the result of incident reporting. This serves the double function of both informing the reporters and congratulating them on their collective contribution to aviation safety.

British Airways Safety Services also disseminate their BASIS information in a variety of ways. In addition to unit reports and journal articles, they issue *Flywise*, an 18–20 page monthly bulletin that includes trend analyses relating to selected events and brief accounts of incidents broken down by fleets. Each incident is assigned both a

SEVERITY **RISK MATRIX**

HIGH	C	B	A (severe)
MEDIUM	D	C	B (high)
LOW	E (minimal)	D (low)	C (medium)
	LOW	MEDIUM	HIGH

PROBABILITY OF RECURRENCE

Figure 9.1   **The British Airways risk management matrix used to evaluate the future risk to the company of the recurrence of an event**
The matrix generates risk categories on a scale from A (severe risk) to E (minimal risk).

risk category (based on a risk matrix—see Figure 9.1) and one of the following action categories:

- *Active investigation*—actions to prevent recurrence not fully understood.
- *Action required*—preventive measures have been identified but not yet implemented.
- *Action monitored*—preventive measures have been implemented and their effects are being monitored.
- *Report monitored*—action taken without need for further investigation by Safety Services. Rates of occurrence are being monitored.

The last factor to be considered here is ease of reporting. The format, length and content of the reporting form or questionnaire are extremely important, as is the context in which respondents are expected to make their report. Privacy and a labour-free returning mechanism are all important incentives—or, to put it the other way round, their absence could be a deterrent. O'Leary and Chappell make the following observations regarding the design of the reporting form:

> If a form is long and requires a great deal of time to complete, reporters are less likely to make the effort. If the form is too short, it is difficult to obtain all the necessary information about the incident. In general, the more specific the questions, the easier it is to complete the questionnaire; however, the information provided will be limited by the choice of the questions. More open questions about the reporter's perceptions, judgements, decisions and actions are not subject to this limitation and give the reporter a greater chance to tell the full story. This method is more effective in gathering all the information about an incident, but takes longer and usually requires more analytic resources within the reporting system.[20]

A certain amount of trial-and-error learning may be necessary before an organization hits upon a format that is best suited both to its purpose and to its potential respondents. In this regard, we can learn from the experience of British Airways Safety Services with their confidential human factors questionnaire. In its initial form, it asked a limited number of very specific questions. Some of the questions sought to establish whether either a slip or a mistake had occurred (see Chapter 4 for the technical descriptions of these unsafe acts) and, if so, what were the contributing factors. The latter were listed below each item and required 'yes/no' responses: in the case of action slips and lapses, they included tiredness, time pressure, lack of stimulation and flight deck ergonomics; in the case of mistakes, they included misleading manuals, misleading displays, insufficient training and crew cooperation.

Even though one of the questions asked what went right, several respondents complained about the negative flavour of the questions overall, and it was realized that this could well deter some potential respondents from completing the questionnaire at all. In addition, the BA analysts were unhappy with the validity of some of the data, since the technical distinctions between slips, lapses and mistakes were not well understood by the flight crew respondents. As a result, BA Safety Services launched a new jargon-free questionnaire in 1995. This asked open-ended questions covering a range of factors from local flight deck influences to the effectiveness of training. O'Leary gives two questions as examples of this new approach:

- How did you and the crew initially respond to the event, and how did you establish what technical and personal issues were involved?
- Was all the relevant flight, FMS and system information clearly available and were all the controls and selectors useful and un-ambiguous? If not, how could these be improved?[21]

As O'Leary observes, this style of questioning moves the analytic workload from the reporter to the human factors analysts. Although the change has increased demand on the limited resources available to process the data, it has made the questionnaire sensitive to a variety of issues not previously covered. In addition, the reliability of the analysis has improved dramatically: 'Previously, some 3500 pilots and engineers may report events idiosyncratically. Now we use a team of only a dozen volunteer flight crew analysts'.[22]

With the multiple choice format of the previous form of the human factors questionnaire, it was relatively simple to convert 'yes/no' responses directly into bar charts. With the second, more open-ended version, the BA analysts had to develop an agreed classification structure in order to identify the issues with human factors significance. They were interested in two major categories: crew performance and the influences upon that behaviour. Crew behaviour was subdivided into error descriptors (action slip or mistake), crew team skills and task specifics, such as automatic or manual handling. Influences too were divided into three groups: organizational factors (training, procedures, commercial pressure and the like), environmental factors (airport facilities, weather conditions and the like) and personal factors (automation, complacency, morale and the like). Future developments are directed at establishing causal links between the various factors—a shift from addressing the 'what?' question to tackling the 'why?' question. The aim here is to identify 'resident pathogens' that may contribute to a variety of different problems on the flight deck.

Figure 9.2 gives an idea of what this development might yield in the way of causal analysis. The figure summarizes a fictional incident involving a rejected takeoff. Here, operational stress was created by a busy airfield and communicating with the company to establish load-sheet figures while taxiing out. The climate on the flight deck was poor. The co-pilot felt overloaded but was not able to communicate this to the captain. He focused on one task at a time and did not cross-check what the captain was doing. As a result, he omitted the 'before takeoff' checks. Takeoff clearance was given as the aircraft approached the runway and, as takeoff power was set, the configuration warning horn indicated that no flap had been selected and the takeoff was aborted.

Finally in this context of engineering a reporting culture, it is worth asking whether there is any scientific evidence to support the efficacy of near-miss accident reporting. In one Swedish study,[23] relating

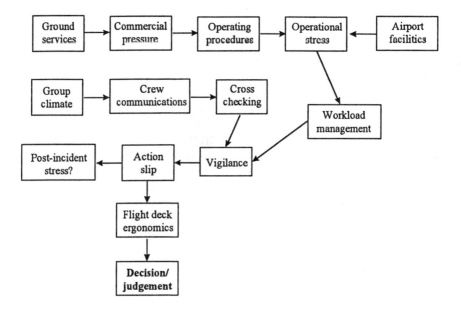

Figure 9.2 Flow diagram of the rejected takeoff incident showing judged causal linkages (after O'Leary)[24]

primarily to individual accidents, the implementation of an incident reporting scheme resulted in an increase in the number of remedial suggestions from the workforce but no significant reduction in the accident rate. In a follow-up study, participants received training in how to recognize and interpret critical incidents. This resulted in a 56 per cent reduction in the severity of injuries but no drop in the accident frequency rate. The main message of these findings is that potential respondents need to be very clear about what constitutes an incident. In some situations, this is not always intuitively obvious.

Engineering a Just Culture

A wholly just culture is almost certainly an unattainable ideal. However, an organization in which the majority of its members share the belief that justice will usually be dispensed is within the bounds of possibility. Two things are clear at the outset. First, it would be quite unacceptable to punish all errors and unsafe acts regardless of their origins and circumstances. Second, it would be equally unacceptable to give a blanket immunity from sanctions to all actions that could, or did, contribute to organizational accidents. While this book has strongly emphasized the situational and systemic factors leading to the catastrophic breakdown of hazardous technologies, it would be naïve not to recognize that, on some relatively rare occasions, accidents can happen as the result of the unreasonably reckless, negligent or even malevolent behaviour of particular individuals. The difficulty lies in discriminating between these few truly 'bad behaviours' and the vast majority of unsafe acts to which the attribution of blame is neither appropriate nor useful.

A prerequisite for engineering a just culture is an agreed set of principles for drawing the line between acceptable and unacceptable actions. To this end, we will start by outlining some of the psychological and legal issues that must be taken into account when making this judgement. Figure 9.3 sets the scene.

All human actions involve three core elements:

- An *intention* that specifies an immediate goal and—where these goal-related actions are not wholly automatic or habitual—the behaviour necessary to achieve it.
- The *actions* triggered by this intention—which may or may not conform to the action plan.
- The *consequences* of these actions—which may or may not achieve the desired objective. The actions can be either successful or unsuccessful in this respect.

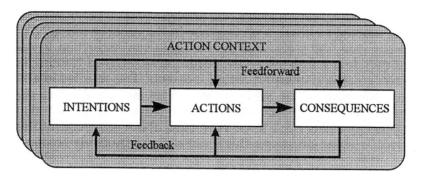

Figure 9.3   The basic elements of human action

The feedforward arrows shown in Figure 9.3 indicate that, in formulating an intention, actions are selected in the belief that they will achieve the goal (or at least provide useful feedback to ensure the success of future actions), but this belief is not always justified. The feedback arrows complete the loop by providing information about the success or otherwise of the preceding actions and their outcomes. Human actions are embedded in a context that contains both the immediate physical environment and the purpose of the behavioural sequence of which a particular action forms a part. The historical context is shown symbolically in Figure 9.3 by the three preceding action frames in the background.

Both of these issues have a close bearing on individual responsibility. In the case of hazardous technologies it is inevitable that all physical situations will contain an element of risk. But is also likely that individual actors will have been—or should have been—trained to foresee and to minimize these risks. This brings us back to the distinction made in Chapter 4 between successful and unsuccessful behaviour on the one hand, and correct and incorrect behaviour on the other. Although success is determined solely by whether the planned actions achieve their immediate objectives, success does not necessarily mean correctness. Successful actions may be incorrect. That is, they could achieve their local purpose and yet be either reckless or negligent.

In the law, a person who acts recklessly is one who takes a deliberate and unjustifiable risk (that is, one that is foreseeable, and where a bad outcome is likely though not certain). However, as Smith and Hogan point out:

> The operator of aircraft, the surgeon performing an operation and the promoter of a tightrope act in the circus must all foresee that their acts might cause death; but we should not describe them as reckless, un-

less the risk taken was unjustifiable. Whether the risk is justifiable depends on the social value of the activity involved, as well as on the probability of the occurrence of the foreseen evil.[25]

Negligence, on the other hand, involves bringing about a consequence that a 'reasonable and prudent' person would have foreseen and avoided. One can also be negligent with regard to a circumstance: 'A person acts negligently with respect to a circumstance when a reasonable man would have been aware of the existence of the circumstance and, because of its existence would have avoided acting in that manner.'[26] In the latter case whether the person failed to foresee the bad outcome and was unaware of the circumstance is irrelevant. For example, X picks up a gun, believing it to be unloaded, points it at Y and pulls the trigger. If any reasonable person would have realized that the gun might possibly be loaded, and thus avoided acting in this way, then X was negligent with regard to circumstance. If the gun was loaded and kills Y, then X was negligent with regard to consequence. In a court of law, it is not necessary for the prosecution to prove anything at all about the person's state of mind at the time of the act. It is enough to establish that particular actions were carried out in certain circumstances. Negligence is historically a civil rather than a criminal law concept, and has a much lower level of culpability than recklessness.[27]

Those involved in the operation of hazardous technologies are often perceived as carrying an additional burden of responsibility by virtue of their training and of the great risks associated with human failure. For example, in the case of *Alidair* v. *Taylor* in 1978, Lord Denning ruled that:

> There are activities in which the degree of professional skill which must be required is so high, and the potential consequences of the smallest departure of that high standard are so serious, that one failure to perform in accordance with those standards is enough to justify dismissal.[28]

This 'hang them all' judgement is unsatisfactory in many respects. It ignores the ubiquity of error as well as the situational factors that promote it. Nor is it sensitive to the varieties of human failure and their differing psychological origins. Pushing this judgement to an absurd conclusion, it could be claimed that, since all pilots, control room operators and others with safety-critical jobs in hazardous technologies are fallible, they will all, at some time or another, inevitably fall short of Lord Denning's 'high standards' and so should all be sacked. Even wise and distinguished judges do not get it right all of the time.

A much sounder guideline is Neil Johnston's *substitution test.*[29] This is in keeping with the principle that the best people can make the worst errors. When faced with an accident or serious incident in which the unsafe acts of a particular person were implicated, we should perform the following mental test. Substitute the individual concerned for someone else coming from the same domain of activity and possessing comparable qualifications and experience. Then ask the following question: 'In the light of how events unfolded and were perceived by those involved in real time, is it likely that this new individual would have behaved any differently?' If the answer is 'probably not' then, as Johnston put it, '... apportioning blame has no material role to play, other than to obscure systemic deficiencies and to blame one of the victims'. A useful addition to the substitution test is to ask of the individual's peers: 'Given the circumstances that prevailed at that time, could you be sure that you would not have committed the same or similar type of unsafe act?' If the answer again is 'probably not', then blame is inappropriate.

So much for the background. We will now turn to the task of grading unsafe acts according to their blameworthiness. In this, as in jurisprudence, a crucial discriminator is the nature of the intention. A crime has two key elements: the *mens rea,* or 'guilty mind' and the *actus reus,* or 'guilty act'. Both are necessary for its commission. Except in very specific instances (as, for example, in the case of negligence), the act without the mental element is not a crime. While, for the most part, we are not concerned here with criminal behaviour, we will adopt the 'mental element' principle as a basic guideline. But hereafter we will approach the issue more from a psychological perspective than from a legal one.

Figure 9.4 sketches out the bare essentials of a decision tree for discriminating the culpability of an unsafe act. It is assumed that the actions under scrutiny have contributed either to an accident or to a serious incident in which a bad outcome was only just averted. In an organizational accident, there are likely to be a number of different unsafe acts, and the decision tree is intended to be applied separately to each of them. Our concern here is with individual unsafe acts committed by either a single person or by different people at various points in the accident sequence.

The key questions relate to intention. If both the actions and the consequences were intended, then we are likely to be in the realm of criminal behaviour and that is probably beyond the scope of the organization to deal with internally. Unintended actions define slips and lapses—in general, the least blameworthy of errors—while unintended consequences cover mistakes and violations. The decision tree usually treats the various error types in the same way, except with regard to the violations question. For mistakes, the question

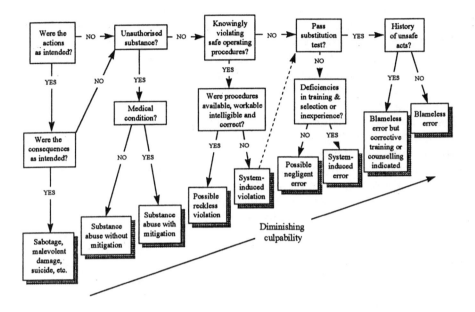

Figure 9.4   A decision tree for determining the culpability of unsafe acts

reads as shown in Figure 9.4, but for slips and lapses, the question relates to what the person was doing when the slip or lapse occurred. If the individual was knowingly engaged in violating safe operating procedures at that time, then the resulting error is more culpable since it should have been realized that violating increases both the likelihood of making an error and the chances of bad consequences resulting (see Chapter 4).

The 'unauthorized substance' question seeks to establish whether or not the individual was under the influence of alcohol or drugs known to impair performance at the time the unsafe act was committed. Since the ingestion of unauthorized substances is usually a voluntary act, their involvement would indicate a high level of culpability. But the matter is not entirely straightforward. In 1975, during a descent towards Nairobi, the co-pilot of a Boeing 747 misheard an air traffic control instruction. Instead of 'seven five zero zero', he heard 'five zero zero zero' and set the autopilot to level out at 5000 feet.[30] Unfortunately, that would have placed the aircraft in a tunnelling mode since it was around 300 feet below the unusually high airfield. When the aircraft broke cloud, the flight crew saw the ground a little more than 200 feet below them. Prompt action by the captain prevented this from being the first major disaster involving a Boeing 'jumbo' jet. It later transpired that the co-pilot had picked up a large tapeworm on a holiday in India and was dosing himself with un-authorized drugs that had, among their side-effects, drowsiness and nausea. Taking unauthorized medication as the result of a medical condition, while clearly reprehensible, is less blameworthy than tak-ing drugs or alcohol for 'recreational purposes' and, as such, in Figure 9.4 it has been assigned to the category of 'substance abuse with mitigation'. The degree of mitigation will, of course, depend upon the local circumstances.

Except when non-compliance has become a largely automatic way of working (as sometimes happens in the case of routine short-cuts), violations involve a conscious decision on the part of the perpetrator to break or bend the rules. However, while the actions may be delib-erate, the possible bad consequences are not—in contrast to sabotage in which both the act and the consequences are intended. Most viola-tions will be non-malevolent in terms of intent, so the degree to which they are blameworthy will depend largely on the quality and availability of the relevant procedures. These, as discussed in Chap-ter 4, are not always appropriate for the particular situation. Where this is judged to be the case—perhaps by a 'jury' of the perpetrator's peers—the problem lies more with the system than with the indi-vidual. However, when good procedures were readily accessible but deliberately violated, the question must arise as to whether the be-haviour was reckless in the legal sense of the term. Such actions are clearly more culpable than 'necessary' violations—that is, non-com-pliant actions necessary to get the job done when the relevant procedures are either wrong, inappropriate or unworkable.

It seems appropriate to apply Johnston's substitution test once the issues of possible substance abuse and deliberate non-compliance have been settled, although something like it clearly has a part to play in judging the culpability of system-induced violations (as indi-cated by the dotted arrow in Figure 9.4). The issue is reasonably straightforward. Could (or has) some well motivated, equally com-petent and comparably qualified individual make (or made) the same kind of error under those or very similar circumstances? If the answer given by a 'jury' of peers is 'yes', then the error is probably blameless. If the answer is 'no', then we have to consider whether there were any system-induced deficiencies in the person's training, selection or experience. If such latent conditions are not identified, then the possibility of a negligent error must be considered. If they are found, it is likely that the unsafe act was a largely blameless system-induced error.

Such a category would apply to the technician whose miswiring of a signal box significantly contributed to the Clapham Junction rail disaster (see Chapter 5). His actions would not pass the substitution test, since he was largely self-taught and had acquired his bad work practices in the absence of adequate training and supervision. But, as the Inquiry established, the underlying problems were those of the system rather than the individual, who was hardworking and highly motivated to do a good job.

In legal jargon, the last major question at the top right-hand corner of Figure 9.4 could be rephrased as 'Any previous?'. People vary widely and consistently in their liability to everyday slips and lapses. For example, some individuals are considerably more absentminded than others. If the person in question has a previous history of unsafe acts, it does not necessarily bear upon the culpability of the error committed on this particular occasion, but it does indicate the necessity for corrective training or even career counselling along the lines of 'Don't you think you would be doing everyone a favour if you considered taking on some other job within the company?'. This is the way that management acquires some of its most distinguished members. Absentmindedness has nothing whatsoever to do with ability or intelligence, but it is not a particularly helpful trait in a pilot or control room operator.

So where should the line be drawn on Figure 9.4 between acceptable and unacceptable behaviour? The most obvious point would be between the two substance abuse categories. Both malevolent damage and the dangerous use of alcohol or drugs are wholly unacceptable and should receive very severe sanctions, possibly administered by the courts rather than the organization. Between 'substance abuse with mitigation' and 'possible negligent error' lies a grey area in which careful judgement must be exercised. The remaining categories should be thought of as blameless—unless they involve aggravating factors not considered here. Experience suggests that the majority of unsafe acts—perhaps 90 per cent or more—fall into this blameless category.

What should happen to the small proportion of individuals whose unsafe acts are justly considered culpable? It is not within the competence of this chapter to advise on the nature of the sanctions. Although this is a matter for the organizations concerned, we can say something about the value—or otherwise—of punishments.

Unfortunately, a large amount of psychological research concerned with the issues of reward and punishment has involved the white rat, and is not especially relevant. Figure 9.5 summarizes in a very simplified way what psychologists know about the effects of reward and punishment in the workplace.[31] The principal issue here is the effectiveness of 'sticks and carrots' in enhancing the likelihood of

	Immediate	*Delayed*
Reward	Positive effects	Doubtful effects
Punishment	Doubtful effects	Negative effects

Figure 9.5 Summary of the effects of reward and punishment on behavioural change in the workplace

desired behaviour and reducing the chances of unwanted behaviour. Rewards are the most powerful means of changing behaviour, but they are only effective if delivered close in time and place to the behaviour that is desired. Delayed punishments have negative effects: they generally do not lead to improved behaviour and can induce resentment in both the punished and the could-be-punished. The cells labelled 'doubtful effects' mean that, in each case, there are opposing forces at work. Hence, the results are uncertain.

But there are other factors that argue strongly in favour of punishing the few who commit egregious unsafe acts. In most organizations the people in the front line know very well who the 'cowboys' and the habitual rule-benders are. Seeing them get away with it on a daily basis does little for morale or for the credibility of the disciplinary system. Watching them getting their 'come-uppance' is not only satisfying, it also serves to reinforce where the boundaries of acceptable behaviour lie. Moreover, outsiders are not the only potential victims. Justified dismissal protects the offender's colleagues. Perhaps more than other possible victims, they are likely to be endangered by the person's repeated recklessness or negligence. Their departure makes the work environment a safer place and also encourages the workforce to perceive the organizational culture as just. Justice works two ways. Severe sanctions for the few can protect the innocence of the many.

David Marx, an aircraft engineer who was one of the principal architects of the Boeing's Maintenance Error Decision Aid (see Chapter 7), made the following comments on the relationship between reporting and disciplinary systems. Though he is writing about the aviation industry, the points are widely applicable:

Many of us have found today's disciplinary systems to be a significant obstacle to asking an employee to come forward and talk about his or her mistake. Consequently, as an industry, we have begun to re-evaluate the inter-relationship of employee discipline and event investigation. Many programs have been developed, both internal to an airline and in association with the FAA ... Whether it is called immunity, amnesty or 'performance-related incentive'—each program attempts to encourage the erring employee to come forward. Yet, as more incentive programs enter the marketplace of ideas, the disciplinary landscape becomes increasingly complex and confusing. With all the programs today, the individual employee needs to be a lawyer to assess whether it is safe to come forward.[32]

David Marx has recently taken a law degree and one of the most interesting products of this marriage between engineering and the law has been the computerized incident investigator, the Aurora Mishap Management System (AMMS). AMMS has a number of elements. For our present purposes, its most important aspect is a structured methodology for establishing the applicability of disciplinary action. This investigative tool is used by an organization's disciplinary review board to aid their decision-making. It applies a common and consistent approach to the issue of determining whether or not disciplinary action is warranted. To date, it has been used in the field of aircraft maintenance by a number of US airlines and has the backing of the Machinists Union.

Engineering a Flexible Culture

Organizational flexibility means possessing a culture capable of adapting effectively to changing demands. Flexibility is one of the defining properties of what a highly influential Berkeley research group—led by Todd La Porte, Karlene Roberts and Gene Rochlin—have termed *high-reliability organizations* (HROs). The group has conducted field research in a number of highly complex, technology-intensive organizations that must operate, as far as humanly possible, to a failure-free standard. The systems of interest here are air traffic control and naval air operations at sea.

The operational challenges facing these (and comparable) organizations are twofold:

- to manage complex, demanding technologies, making sure to avoid major failures that could cripple, perhaps destroy, the organization;
- at the same time, to maintain the capacity for meeting periods of very high, peak demand and production whenever these occur.[33]

The organizations studied by the Berkeley group had the following characteristics:

- They were large, internally dynamic and intermittently intensely interactive.
- Each performed complex and exacting tasks under considerable time pressure.
- They have carried out these demanding activities with a very low error rate and an almost complete absence of catastrophic failure over a number of years.

On the face of it, both of the organizations to be considered here—the US Navy nuclear aircraft carrier and the air traffic control centre—had highly bureaucratic and hierarchical organizational structures, each with a clear line of authority and command. Both organizations relied heavily on tested standard operating procedures (SOPs). Both organizations invested a great deal of effort in training people in the use of these procedures. It was almost the case that, under routine operating conditions, the only decision necessary was which SOP to apply.

Actions in these HROs were closely monitored so that immediate investigations—termed 'hot washups' in the US Navy—were conducted whenever errors occurred. Over the years, these organizations have learned that there are particular kinds of error, often quite minor, that can escalate rapidly into major, system-threatening failures. Trial-and-error learning in these critical areas was not encouraged, as it was elsewhere, in case it should become 'habit-forming.' Also, as La Porte and Consolini describe it: 'there is a palpable sense that there are likely to be similar events that cannot be foreseen clearly, and that may be beyond imagining. This is an ever-present cloud over operations, a constant concern'.[34] In short, these organizations suffer chronic unease. The following quotation from the same source captures this intelligent wariness and its cultural consequences very eloquently:

> The people in these organizations know almost everything technical about what they are doing—and fear being lulled into supposing that they have prepared for any contingency. Yet even a minute failure of intelligence, a bit of uncertainty, can trigger disaster. They are driven to use a proactive, preventative decision making strategy. Analysis and search come before as well as after errors. They try to be synoptic while knowing that they can never fully achieve it. In the attempt to avoid the pitfalls in this struggle, decision making patterns appear to support apparently contradictory production-enhancing and error-reduction strategies. The patterns encourage

- reporting errors without encouraging a lax attitude toward the commission of errors;
- initiative to identify flaws in SOPs and nominate and validate changes in those that prove to be inadequate;
- error avoidance without stifling initiative or (creating) operator rigidity; and
- mutual monitoring without counter-productive loss of operator confidence, autonomy and trust.[35]

So how do HROs respond to bursts of high-tempo operations? Lying in wait beneath the surface of the routine, bureaucratic, SOP-driven mode is quite another pattern of organizational behaviour. Here is what happened aboard the aircraft carrier when some 70 of its 90 aircraft were flying off on missions:

> Authority patterns shift to a basis of functional skill. Collegial authority (and decision patterns) overlay bureaucratic ones as the tempo of operations increases. Formal rank and status decline as a reason for obedience. Hierarchical rank defers to technical expertise often held by those of lower formal rank. Chiefs (senior non-commissioned officers) advise commanders, gently direct lieutenants and cow ensigns. Criticality, hazards, and sophistication of operations prompt a kind of functional discipline, a professionalization of the work teams. Feedback and (sometimes conflictual) negotiations increase in importance; feedback about 'how goes it' is sought and valued.[36]

A similar kind of flexibility was evident in the air traffic control centre. Sudden wind shifts can impose a high additional burden on already busy controllers. Reorienting the flight paths of a large number of aircraft in relation to what, in this instance, were three major airports, two large military airbases and five smaller general aviation airfields becomes a major programme for the controllers on duty. La Porte and Consolini described what happened:

> The tempo at the approach-control facility and the enroute center increases, and controllers gather in small groups around relevant radar screens, plotting the optimal ways to manage the traffic as the shift in [wind] direction becomes imminent. Advice is traded, suggestions put forward, and the actual traffic is compared with the simulations used in the long hours of training the controllers undergo.... . While there are general rules and controllers and supervisors have formal authority, it is the team that rallies round the controllers in 'the hot seats'. It will be the experienced controller virtuosos [rather than the supervisors] who dominate the decision train. 'Losing separation'—the key indicator of controller failure—is too awful to trust to rules alone.[37]

When the high-tempo period slackens off, authority reverts seamlessly to its previous bureaucratic, rank-determined form. A very similar type of flexibility was evident in an anecdote which I came across concerning one of the most highly rated US Army units of the Korean War. The senior NCOs of the unit recognized that they lacked the qualities to lead men in action. When the unit went into combat, local command passed to a small group of enlisted men. Afterwards, these 'combat leaders' were quite happy to follow the orders of the NCOs, whose skills in everyday soldiering they fully recognized.

There is, then, convincing evidence that an organization's ability to switch from a bureaucratic, centralized mode to a more decentralized professional mode is an important determinant of reliability—or even survival. But how can it be engineered? Karl Weick—whose work has been cited at various points throughout this book—has made a number of important observations in this regard. In order to achieve effective decentralization—of the kind described earlier— Weick argues that:

> ... you first have to centralise so that people are socialised to use similar decision premises and assumptions so that when they operate their own units, these decentralised operations are equivalent and co-ordinated. This is precisely what culture does. It creates a homogeneous set of assumptions and decision premises which, when they are invoked on a local and decentralised basis, preserve co-ordination and centralisation. More important, when centralisation occurs via decision premises and assumptions, compliance occurs without surveillance. This is in sharp contrast to centralisation by rules and regulations or centralisation by standardisation and hierarchy, both of which require high surveillance. Furthermore, neither rules nor standardisation are well equipped to deal with emergencies for which there is no precedent.[38]

It is probably no coincidence that the HROs studied by the Berkeley group were either military or had many key personnel with a military background—this applies equally to the third HRO not discussed above, a Californian nuclear power plant in which many operators and supervisors had been in the nuclear Navy. The acceptance of a disciplined approach to working, well founded trust in SOPs, and a familiarity with the ways of rank-based structures would all help to forge the shared values about reliability that permit effective decentralized action when the occasion demands.

Weick makes another point of considerable relevance here. All hazardous technologies face the problem of requisite variety—the variety that exists in the system exceeds the variety of the people who must control it (see Chapter 4). As a result, 'they miss important information, their diagnoses are incomplete, and their remedies are short-sighted and can magnify rather than reduce a problem'. But this problem, can be reduced by a culture that encourages 'war stories'. Since the nature of these systems allows little scope for trial-and-error learning, maintaining reliability depends on developing alternatives for trial and error. These could include imagination, vicarious experience, simulation, stories and story-telling.

> A system that values stories and storytelling is potentially more reliable because people know more about their system, know more of the potential errors that might occur, and they are more confident that they can handle those errors that do occur because they know that other people have already handled similar errors.[39]

Other ways of reducing the gap between the variety of the system and the variety of its human controllers include:

- *A culture that favours face-to-face communication.* 'One way to describe (admittedly stereotype) engineers is as smart people who don't talk. Since we know that people tend to devalue what they don't do well, if high reliability systems need rich, dense talk to maintain complexity, then they may find it hard to generate this richness if talk is devalued or if people are unable to find substitutes for talk (e.g., electronic mail may be a substitute).'[40]
- *Work groups made up of divergent people.* 'A team of divergent individuals has more requisite variety than a team of homogeneous individuals.'[41] It matters less what makes up this diversity—different speciality, different experience, different gender, and the like—than the fact that it exists. 'If people look for different things, when their observations are pooled they collectively see more than any one of them alone would see.'[42] By the same token, groups made up of very similar people tend to see very similar things, and so lack requisite variety.

The decentralization of authority under certain conditions was a crucial feature of the German military concept of *Auftragssystem*— (mission system) discussed in Chapter 4. Its essence was that a subordinate commander, a subaltern or senior NCO, should be trained to a level where he (or, very rarely, she) could achieve the tactical goals of superior officers, with or without orders. Translating this into a civilian context, it means selecting and training first-line supervisors so that they are able to direct safe and productive working without the need for SOPs. Such a localized system of behavioural guidance makes heavy demands on the personal qualities of the supervisors. A prerequisite is an extensive experience of the jobs carried out in the workplace and the conditions under which they are likely to be performed. Supervisors need to be 'sitewise' both to the local productive demands and to the range of obvious and less obvious hazards. Equally important is a personal authority derived both from the respect of the workforce and the support of management—a key feature in the success of the German Army.

Not all activities in hazardous technologies are carried out in supervised groups. When people are relatively isolated, the onus shifts from group to self-controls. Crucial among these are the techniques designed to enhance hazard awareness and risk perception, These are the measures that seek to promote 'correct' rather than merely 'successful' performance. A number of hazard evaluation programmes are being developed or have already been implemented. However, as Willem Albert Wagenaar has observed,[43] risk appraisal training is of little value once the incorrect actions have become habitual. When this happens, people are not *taking risks* deliberately, they are *running risks* in a largely thoughtless and automatic fashion. To be effective, such training must occur in the initial phase of employment and then be consolidated and extended by on-the-spot supervisory guidance. By the same token, it is mainly through local supervisory interventions that long-established pattern of incorrect behaviour can be modified.

In summary, high-reliability organizations are able to shift from centralized control to a decentralized mode in which the guidance of local operations depends largely upon the professionalism of first-line supervisors. Paradoxically perhaps, the success of this transformation depends on the prior establishment of a strong and disciplined hierarchical culture. It is the shared values and assumptions created by this culture that permit the coordination of decentralized work groups. Effective teams, capable of operating autonomously when the circumstances demand it, need high-quality leaders. This, in turn, requires that the organization invest heavily in the quality, motivation and experience of its first-line supervisors.

Engineering a Learning Culture

Of all the 'subcultures' so far considered, a learning culture is probably the easiest to engineer but the most difficult to make work. Most of its constituent elements have already been described—observing (noticing, attending, heeding, tracking), reflecting (analysing, interpreting, diagnosing), creating (imagining, designing, planning) and acting (implementing, doing, testing). The first three are not so difficult. It is the last one—acting—that is likely to cause most of the problems. Echoing the rueful remark by the man from Barings Bank after the collapse—there always seemed to be something more pressing to do.

Beyond what has already been written,[44] there is little more that a book can do to give top managers the will to put in place the reforms indicated by their safety information systems, except to bring to their attention the chilling observation of the organizational theorist, Peter Senge:

> Learning disabilities are tragic in children, but they are fatal in organizations. Because of them, few corporations live even half as long as the person—Most die before they reach the age of forty.[45]

Senior managers should not need to be reminded that an organizational accident can brutally cut short even that brief span.

Safety Culture: Far More than the Sum of its Parts

At this point, I have in mind an imaginary technical manager from an organization with a good safety record (probably measured in LTIFs) who starts to count off the cultural elements that have so far been considered. Yes, he or she might decide, we have an incident reporting system of sorts. Yes, we have a reasonably fair and straightforward method of deciding whether or not disciplinary action is warranted. Yes, we have, on occasions, allowed our first-line supervisors a good deal of latitude and backed up their decisions afterwards—when things turn out all right, of course. And, yes, we have implemented a number of fairly expensive safety improvements on the basis of both reactive and proactive information, so it could be said that we have a learning culture. Does all of this mean that we have an informed culture—or, in more usual terms, a safety culture?

As any engineer knows, assembling the parts of a machine is not the same thing as making it work. And the same is even more true of social engineering than of its more mechanical counterparts. In order to answer our hypothetical manager, we would have to pose some questions in return:

- Which board members have responsibility for organizational safety—as opposed to conventional health and safety at work concerns?
- Is information relating to organizational safety discussed at all regular board meetings—or their high-level equivalent?
- What system, if any, do you have for costing the losses caused by unsafe acts, incidents and accidents?

- Who collates, analyses and disseminates information relating to organizational safety? By how many reporting levels is this individual separated from the CEO? What annual budget does this person's department receive? How many staff does he or she oversee?
- Is a safety-related appointment seen as rewarding talent (a good career move) or as an organizational oubliette for spent forces?
- How many specialists in human and organizational factors does the company employ?
- Who decides what disciplinary action should be meted out? Are the 'defendant's' peers and union representatives involved in the judgement process? Is there any internal appeals procedure?

The potential list is endless. The point is this—the mere possession of the 'engineered' externals is not enough. A safety culture is far more than the sum of its component parts. And here—perversely perhaps, considering what was said at the beginning of the chapter—we must acknowledge the force of the argument asserting that a culture is something that an organization 'is' rather than something it 'has'. But if it is to achieve anything approaching a satisfactory 'is' state, it first has to 'have' the essential components. And these, as we have tried to show, can be engineered. The rest is up to the organizational chemistry. But using and doing—particularly in a technical organization—lead to thinking and believing.

Finally, it is worth pointing out that if you are convinced that your organization has a good safety culture, you are almost certainly mistaken. Like a state of grace, a safety culture is something that is striven for but rarely attained. As in religion, the process is more important than the product. The virtue—and the reward—lies in the struggle rather than the outcome.

Postscript: National Culture

Every organizational culture is shaped by the national context in which it exists—and this is especially true for multinational organizations. It is not within the scope of this chapter to deal with the differences in national culture. For this, the reader is directed to the seminal books by Geert Hofstede.[46] The interested reader is also strongly advised to seek out the work of Robert Helmreich[47] and his colleagues at the University of Texas, and of Najmedin Meshkati[48] at the University of Southern California.

Notes

1   Data from the Flight Safety Foundation Icarus Committee, cited by Skandia International. I am grateful to Lars Högberg of the Swedish Nuclear Power Inspectorate (SKI) for sending me this information.

2   B. Uttal, 'The corporate culture vultures', *Fortune*, 17 October 1983.

3   T.E. Deal and A.A. Kennedy, *Corporate Cultures: The Rites and Rituals of Corporate Life*, (Reading, MA: Addison-Wesley, 1982).

4   T.J. Peters and R.H. Waterman, *In Search of Excellence: Lessons from America's Best-Run Companies*, (New York: Harper & Row, 1982).

5   Ibid.

6   Ibid., p. 76.

7   P. Bate, 'The impact of organizational culture on approaches to organizational problem-solving', in G. Salaman (ed.), *Human Resource Strategies*, (London: Sage, 1992), p. 229 cited by N. Thompson, S. Stradling, M. Murphy and P. O'Neill, 'Stress and organizational culture', *British Journal of Social Work*, **26**, 1996, pp. 647–65.

8   Cited by Thompson *et al.*, op. cit. p. 651.

9   G. Hofstede, *Cultures and Organizations: Intercultural Cooperation and its Importance for Survival*, (London: Harper Collins, 1994), p. 199.

10 International Nuclear Safety Advisory Group (IAEA), *Safety Culture*, (Vienna: IAEA, 1991).

11 Cited by R. Booth, 'Safety culture: concept, measurement and training implications', *Proceedings of British Health and Safety Society Spring Conference: Safety Culture and the Management of Risk*, 19–20 April, 1993, p. 5.

12 M. O'Leary and S.L. Chappell, 'Confidential incident reporting systems create vital awareness of safety problems', *ICAO Journal*, **51**, 1996, pp. 11–13. Dr Chappell is now Program Manager, Human Factors Services, at TransQuest Inc. in Atlanta, Georgia.

13 Ibid., p. 11.

14 S.L. Chappell, 'Aviation Safety Reporting System: program overview' in *Report of the Seventh ICAO Flight Safety and Human Factors Regional Seminar*, Addis Ababa, Ethiopia, 18–21 October, 1994, pp. 312–53.

15 J.A. Passmore, 'Air safety report form', *Flight Deck*, Spring 1995, pp. 3–4.

16 M. O'Leary and N. Pidgeon, 'Too bad we have to have confidential reporting programmes', *Flight Deck*, Summer 1995, pp. 11–16.

17 M. O'Leary and S. Fisher, *British Airways Confidential Human Factors Reporting Programme. First Year Report, April 1992–March 1993*, (Hounslow: British Airways Safety Services, 1993).

18 S. Chappell, op. cit.

19 M. O'Leary and S. Chappell, op. cit.

20 Ibid., p. 12.

21 M. O'Leary, 'New developments in the British Airways Confidential Human Factors Reporting Programme', *Flight Deck*, Summer 1996, pp. 19–22.

22 Ibid., p.21.

23 Cited by S.J. Guastello, 'Do we really know how well our occupational accident prevention programs work?', *Safety Science*, **16**, 1993, pp. 445–63.

24 O'Leary, op. cit. 1996, p. 22.

25 J.C. Smith and B. Hogan, *Criminal Law*, (3rd edn), (London: Butterworths, 1975), p. 45.

26 Ibid., pp. 45–6.

27 D. Marx, Personal communication, 9 January 1997. I am most grateful to David

Marx for his helpful advice on the distinction between recklessness and negligence.

28 Cited by M. O'Leary and N. Pidgeon, op. cit., p. 16.

29 N. Johnston, 'Do blame and punishment have a role in organizational risk management?', *Flight Deck*, Spring 1995, pp. 33–6.

30 Air Accident Investigation Branch (AAIB), *Boeing 747-136 G-AWNJ. Report on the Incident near Nairobi Airport, Kenya on 3 September 1974*, Aircraft Accident Report 14/75, (London: HMSO, 1975).

31 J.M. George, 'Asymmetrical effects of rewards and punishment: the case of social loafing', *Journal of Occupational and Organizational Psychology*, 68, 1995, pp. 327–28.

32 D. Marx, Personal communication, 1 October 1996.

33 T.R. LaPorte and P.M. Consolini, 'Working in practice but not in theory: theoretical challenges of "high-reliability" organizations', *Journal of Public Administration Research and Theory*, 1, 1991, p. 21.

34 Ibid., p. 27.

35 Ibid., p. 29.

36 Ibid., p. 32.

37 Ibid., p. 34.

38 K.E. Weick, 'Organizational culture as a source of high reliability', *California Management Review*, 24, 1987, pp. 112–27, see p. 124.

39 Ibid., p. 113.

40 Ibid., p. 115. See also K.M. Eisenhardt, J.L. Kahwajy and L.J. Bourgeois, 'Conflict and strategic choice: How top management teams disagree', *California Management Review*, 39, 1997, pp. 42–62.

41 K.E. Weick, op. cit., 1987, p. 116.

42 Ibid., p. 116.

43 W.A. Wagenaar, 'Risk-taking and accident causation' in J. Yates (ed.), *Risk-Taking Behaviour*, (Chichester: Wiley, 1992).

44 The interested reader is recommended to start with: B. Toft and S. Reynolds, *Learning from Disaster: A Management Approach*, (London: Butterworths Heinemann, 1994).

45 P.M. Senge, *The Fifth Discipline: The Art and Practice of the Learning Organization*, (London: Century Business, 1990).

46 G. Hofstede, op. cit., 1994. See also G. Hostede, *Culture's Consequences: International Differences in Work-Related Values*, (Beverly Hills, CA: Sage Publications, 1980).

47 R.L. Helmreich, A. Merritt and P. Sherman, 'Research project evaluates the effect of national culture on flight crew behaviour', *ICAO Journal*, 51, 1996, pp. 14–16 (and many other sources).

48 N. Meshkati, 'Cultural factors influencing safety need to be addressed in design and operation of technology', *ICAO Journal*, 51, 1996, pp. 17–18 (and many other sources).

10 Reconciling the Different Approaches to Safety Management

Revisiting the Distinction Between Individual and Organizational Accidents

Having recently tried out some of the book's ideas on safety professionals dealing with the day-to-day realities of North Sea oil exploration and production,[1] I am conscious that there is something of a gulf between their current focus on personal injury accidents and my emphasis upon the larger-scale, but comparatively rare, organizational accidents. The professionals are, of course, fully aware of the commercial, human and environmental dangers posed by catastrophes like *Piper Alpha*, but, for the last 15 years or so, the principal metric for assessing safety in the oil industry, as in many other hazardous domains, has been LTIF—lost-time injuries per million man hours.[2] It therefore seems appropriate, in this final chapter, to reopen the issue of the differences between individual and organizational accidents that were discussed only briefly in Chapter 1.

Another problem that needs to be confronted is the belief held by many technical managers that the main threat to the integrity of their assets is posed by the behavioural and motivational shortcomings of those at the 'sharp end'. For them, the oft-repeated statistic that human errors are implicated in some 80–95 per cent of all events generally means that individual human inadequacies and errant actions are the principal causes of all accidents. What they hope for in seeking the help of a human factors specialist is someone or something to 'fix' the psychological origins of these deviant and unwanted behaviours. But this—as I hope is now clear—runs counter to the main message of this book. Workplaces and organizations are easier to manage than the minds of individual workers. You cannot change the human condition, but you can change the conditions under which people work. In short, the solutions to

223

most human performance problems are technical rather than psychological.

One way to begin to resolve these apparent conflicts is to recognize that there are three distinct models for managing safety—the *person* model, the *engineering* model and the *organizational* model—and that each of them has a different perspective on human error. These important distinctions were first made by Deborah Lucas,[3] now with the UK Health and Safety Executive.

The principal aim of this concluding chapter is to show that, despite their differences in tradition, emphasis and domains of application, there is no reason why these various models and their associated practices should not coexist harmoniously within the same organization—so long as the strengths and weaknesses of each approach are recognized.

Three Approaches to Safety Management

The Person Model

The person model is exemplified by the traditional occupational safety approach. The main emphases are upon individual unsafe acts and personal injury accidents. It views people as free agents capable of choosing between safe and unsafe behaviour. This means that errors are perceived as being shaped predominantly by psychological factors such as inattention, forgetfulness, poor motivation, carelessness, lack of knowledge skills and experience, negligence and—on occasions—culpable recklessness. Its principal applications are in those domains involving close encounters with hazards. As such, it is the most widely adopted of the three models. It is also the approach with the longest history, stretching back to the beginnings of industrialisation. It is usually policed by safety departments and safety professionals, though—more recently—the accent has been upon personal responsibility.

The most widely used countermeasures are 'fear appeal' poster campaigns, rewards and punishments, unsafe act auditing, writing another procedure, training and selection. Progress is measured by personal injury statistics, such as fatalities, lost-time injuries, medical treatment cases, first aid cases, and the like. It is frequently underpinned by the 'iceberg' or 'pyramid' views of accident causation. The empirical basis for such beliefs was provided by Frank Bird's analysis of 1 753 498 accidents reported by 297 companies, representing 21 different industries.[4] This yielded the now widely used 1:10:30:600 ratio (see below), though other comparable ratios are also employed:

- 1 serious or major injury
- 10 minor injuries
- 30 property damage accidents
- 600 incidents with no visible damage or injury.

The Engineering Model

The engineering model has its origins in reliability engineering, traditional ergonomics (and its modern variant—cognitive engineering) risk management and human reliability assessment. Safety is viewed as something that needs to be 'engineered' into the system and, where possible, to be quantified as precisely as possible. Thus, the focus is upon engineered system reliability, often expressed in probabilistic terms. In contrast to the person model, human errors are not regarded simply as the product of what goes on between an individual's ears. Rather, they emerge from human–machine mismatches, or poor *human engineering*—that is, the failure on the part of the system designers to tailor the system appropriately to the cognitive strengths and weaknesses of its human controllers. Typically, the model focuses on how the performance of front-line operators (for example, control room operators and pilots) is influenced by the characteristics of the workplace or, more specifically, by the informational properties of the human–machine interface. These issues were discussed at some length in Chapter 3 in the context of 'clumsy automation'.

Research in this area was originally supported by the nuclear power industry, the military, the space agencies, the chemical process industry and aviation—domains in which the safety of a system hinges critically on the reliability of a small number of human controllers. More recently, however, the requirement upon oil and gas companies to produce formal safety assessments as part of their safety cases (see Chapter 7) has greatly extended its area of application. The practical applications of this approach include: hazard operability studies (HAZOPS), hazard analysis studies (HAZANS), probabilistic risk assessment (PRA), technical safety audits, reliability and maintainability studies (RAMS), human reliability assessment (HRA), cognitive task analyses, ergonomic guidelines, databases, and the application of decision support systems. Excellent accounts of the nature and application of these tools can be found in a number of recent texts.[5]

The Organizational Model

If the organizational model, the newest of the three, has a disciplinary link, then it would probably be with crisis management. Although not always apparent to its practitioners, it owes its intellectual origins to two books. The first was *Man-Made Disaster* by the late (and greatly missed) Barry Turner, published in 1978.[6] The second major influence was Charles Perrow's *Normal Accidents*.[7] In retrospect, credit must also go to the Hon. Peter Mahon for his remarkable report upon the Mt. Erebus tragedy that occurred in 1979.[8] As Neil Johnston has pointed out, the Mahon Report was ten years ahead of its time.[9] Most of the accidents that have shaped our current thinking about organizational factors had yet to happen. As indicated in Chapter 8, Mr Justice Moshansky's extensive report on the Dryden tragedy has provided a more recent endorsement of the organizational approach.[10]

The organizational model views human error more as a consequence than as a cause. Errors are the symptoms that reveal the presence of latent conditions in the system at large. They are important only in so far as they adversely affect the integrity of the defences. The model emphasises the necessity for proactive measures of 'safety health' and the need for continual reforms of the system's basic processes. As such, it has much in common with Total Quality Management. Indeed, the organizational model deliberately blurs the distinction between safety-related and quality-determining factors. Both are viewed as important for increasing the system's intrinsic resistance to its operational hazards. Both are seen as being implicated in organizational accidents.

In many respects, the organizational model is simply an extension of the engineering model and is in no way incompatible with it. Human–machine mismatches are seen as being the result of prior decisions in the upper echelons of the system. And these, in turn, are shaped by wider regulatory and societal factors.

This book has presented a mixture of the engineering and organizational approaches, with a somewhat greater emphasis on the latter. However, it is quite clear that both are necessary for understanding the aetiology of organizational accidents and for limiting their occurrence. Where there is a conflict, it is between both of these models and the largely person-directed approach of the traditional occupational safety professionals. However, these differences are often more a matter of circumstance than of substance.

Primary Risk Areas

For any hazardous technology there are potentially four primary risk areas. These will, of course, vary in significance from domain to domain.

- *Personal injury or damage risks.* These are associated with activities in which the workforce is in close contact with the hazards. The unwanted outcomes are either injury to the worker or limited damage to some asset. In neither case, however, does this involve extensive damage to an installation or to the system at large. Such events were described in Chapter 1 as individual accidents because they affect either individual workers or individual items of equipment.
- *Risks due to errors committed by key front-line controllers.* These are most closely associated with systems (or subsystems) in which control is centralized in the hands of a relatively few individuals. In modern systems there will almost certainly be some measure of automation involved in the control activity. The unwanted outcome could be an organizational accident though, as indicated earlier, it is unlikely that such an event could arise as the result of a single error on the part of the operator(s).
- *Risks due to the insidious accumulation of latent conditions within the maintenance, managerial and organizational spheres.* Such risks are closely associated with systems possessing several defences-in-depth. As discussed earlier, the unwanted outcome is the breaching or bypassing of critical defences, bringing hazards (which need not necessarily cause physical harm) into damaging contact with people and/or assets to produce losses. Here, an entire installation or system could be destroyed. These are the quintessential organizational accidents.
- *Risks to third parties.* These are risks that threaten the lives, livelihoods and the physical and mental well-being of individuals not directly employed by the organization, as well as the likelihood of losses and damage to their assets or to the environment at large. Such 'third parties' would include passengers, patients, investors, taxpayers, those living in the neighbourhood of a hazardous installation, or indeed anyone adversely affected by the operation of a particular technology, financial activity (for example, banking, insurance, pension fund management and the like) or public service (for example, the military, the police force and the like).

These risk types fall into two groups. Personal injury risks are closely identified with individual accidents and the person model. The remaining three risk types are all associated with organizational accidents and are the main concerns of both the engineering and the organizational models.

The Preponderance of Risks in Different Domains

In some measure at least, all of these risks are present in all hazard-ous operations. In that sense, all three safety management models are applicable. But some risks are more salient than others, and the balance between the risks varies from domain to domain (see Table 10.1). Moreover, this preponderance is not fixed once and for all by the nature of the domain, it can also vary with technological and even societal developments. Advanced manufacturing techniques, for example, have shifted safety concerns from the slips, lapses, trips and fumbles of individual workers on the conventional production line to the costly mistakes of the few key operators who programme computer-driven machine tools. In hospitals, the main worries were once the well-being of patients and the safeguarding of staff from contact with diseases, but societal changes have now forced risk managers to consider the dangers posed by the increasing number of physical assaults upon healthcare workers.

The history of modern technology is rich in instances of risk man-agers being caught with their eyes on the wrong ball. In transport systems, for example, the traditional emphasis has been upon the safety of the passengers or cargo, and on the risks posed by the fallibility of those at the sharp end—the pilots, the train drivers, and the ships' crews. Only relatively recently, for instance, have airlines become aware of the risks associated with maintenance, or of the enormous losses caused by personal injuries to ground-based staff—amounting to $30 million per year in some large US carriers. It took the Clapham Junction disaster to make British Rail aware of the risks associated with technical work on the signalling system. And only in its aftermath did they begin to record the personal injury accidents sustained by their infrastructure staff (for example, shunters and track workers). Yet this was an organization with a 160-year tradition of safety innovation in both the engineering and the human spheres (where human equated to driver or signalman). Only in the last decade has the nuclear power industry started to appreciate the risks associated with low-power and shutdown conditions. The training and procedures for control room operators were almost exclusively geared to handling emergencies in the more typical full-power situation. It took the King's Cross Underground tragedy to reveal that stations, as well as trains, could be dangerous places for passengers and staff. And it took the Cullen Inquiry to make many North Sea oil and gas operators aware of formal safety assessment techniques, though they had long been a staple item on the risk management agenda for the military and for the nuclear power and chemical process industries.

Risk is a function of both the likelihood of an event occurring and of the possible extent of its bad outcome. The estimates given in

Table 10.1 vary widely in the ways in which a high (or very high) risk rating can be achieved. In the personal accident column, for example, the harmful consequences of an event will usually be limited to an individual, to a small group or to their immediate surroundings. Here, high risk ratings are assigned more on the basis of likelihood than outcome—and this, of course, must also depend on the numbers of potential victims. But the reverse is true for, say, the 'very high' rating given to the third-party risks associated with nuclear

Table 10.1   Comparison of estimates of the four risk types across domains

Domain of operation	Personal injury risks	Errors of key operators	Latent conditions	Third-party risks
Nuclear power generation	very low (normal state)	high	high	very high
Chemical process plants	low–moderate (normal state)	high	high	very high
Commercial aviation	moderate–high (ground staff)	high	high	very high
Advanced manufacturing	very low	high	high	variable
Oil exploration and production	high	high	high	very high
Marine	high	high	high	very high
Railways	high (infra-structure)	high	high	high
Construction	very high	moderate–high	high	high
Mining	very high	moderate–high	high	low–moderate
Medicine	moderate	very high	high	very high
Financial services	very low	very high	high	high
Sports stadiums and crowd control	high	high	high	high

power generation. Here, the likelihood of a major release of radioactive materials is exceedingly small but its adverse consequences, in the worst case, are very bad indeed.

Even if readers do not agree with the particular risk estimates given in Table 10.1, one thing remains hard to dispute—there is less variability between domains for the risks associated with latent conditions than for the other risk types. All domains must be assessed as at least 'high' in this regard. The reasons for this are not difficult to find. The further one moves from a domain's front-line operations, the more alike organizations become. Technical systems of whatever kind inevitably share a large number of common processes: forecasting, planning, scheduling, budgeting, specifying new equipment (and sometimes designing and building it), operating, maintaining, managing, communicating and the like. And it is within these processes that the seeds of future disasters are sown—irrespective of the domain. In summary, all hazardous domains are threatened by organizational accidents, but their individual accident risks are extremely variable. Nevertheless, the person model remains the most widely used approach to safety management.

In addition there is a marked asymmetry of application between the person model on the one hand, and the engineering and organizational models on the other. Whereas the latter two approaches can be usefully applied to limiting personal injury risks (and hence preventing individual accidents), the person model is not at all helpful in dealing with key operator error risks, latent conditions or third-party risks—all of which fall squarely into the province of organizational accidents. Indeed, its predominance in the minds of many technical managers is a definite barrier to improved safety. When applied to the appropriate risks, both the person approach and its tools have shown themselves to be valuable. The difficulty lies in the failure on the part of some managers to recognize that there are other types of risk and other tools to deal with them. When all you possess is a hammer, then almost everything looks like a nail. Or, to put it more directly, when the person model is the only approach with which you feel comfortable, then every problem seems to be a person problem.

Why is the person model so seductive? It is worthwhile taking a look at some of the reasons for the widespread appeal of the person-oriented approach to safety management.

- It has been around a long time. Most senior managers grew up with it and are comfortable with its doctrines. For many managers, management equates to 'people management' so the person-oriented approach fits the job description.
- Some organizations, notably Du Pont, have been conspicuously successful in achieving very low LTIF rates. In 1990, for exam-

ple, their worldwide and European lost-time injuries per 200 000 exposure hours (involving more than one day's absence from work) were 0.032 and 0.023, respectively. That is 0.16 and 0.12 per million manhours, or 0.32 and 0.23 per 1000 employees. Du Pont is justifiably seen as the market leader in this regard, and a number of major companies have sought to emulate their achievements, particularly in the domain of oil exploration and production. These numbers provide a very clear target to aim for, and such well defined goals are welcomed in the otherwise rather nebulous business of safety management.

- It is much easier to pin the legal responsibility for an accident on the unsafe acts of those at the 'sharp end'. The connection between these individual actions and the disastrous outcome is far more readily demonstrated than are any possible links between earlier management decisions and the accident—see, for example, the failed prosecution of the managers implicated (by the Sheen Inquiry)[11] into the capsize of the *Herald of Free Enterprise*.[12] In that case, Mr Justice Turner directed the jury to acquit the defendants even before the defence gave evidence. He said that there was no direct evidence that a 'reasonably prudent person' occupying the position of any of the five defendants would have perceived the risk was obvious or serious. It was not enough, he added, to show failures; the defendants had to have been reckless. There was no question, he said, of making a corporation guilty of manslaughter by aggregating the acts of individuals whose actions were not themselves reckless. In this, English law runs counter to a 1988 Council of Europe recommendation that acts of individuals should be accumulated when deciding whether a corporation had committed an offence.
- As mentioned in Chapter 7, we place a high value on personal freedom, or the illusion of free will. Since we also impute this to others, we have a strong tendency to assume that the unsafe acts were committed because the individuals in question chose an unsafe course of action. This, as indicated earlier, is compounded by the *fundamental attribution error*—the universal belief that bad acts are committed by bad people.
- The person model also accords very closely with the way in which people try to establish cause. This is not the place to get embroiled in the philosophy of causation, but it is worth noting what the experts on jurisprudence have had to say on the matter:

> A causal explanation of a particular occurrence is brought to a stop when it has been explained by a deliberate act, in the sense that none of the antecedents will count as the cause of the

occurrence. A deliberate human act is therefore most often a barrier and a goal in tracing back causes ... it is often something through which we do not trace the cause of a later event and something to which we do trace the cause through intervening causes of other kinds.[13]

- Finally, there are two other factors that help to clinch the primacy of the person model in many people's minds. At an individual level, we gain a good deal of emotional satisfaction from blaming someone—rather than something—when things go wrong. And, at the organizational level, there are obvious financial and legal benefits in being able to uncouple individual fallibility from corporate liability. Either way, there are advantages in being able to limit culpability to specific people.

Can Personal Injuries Predict Organizational Accidents?

To remain true to the basic arguments expressed in this book, we must readily acknowledge that both individual and organizational accidents have their roots in upstream organizational and managerial factors. Both types of event are due to latent conditions.

It is not hard to find examples of personal injuries having organizational causes. In his excellent book, *An Engineer's View of Human Error*, Trevor Kletz[14] divides his chapters into two groups—those in which personal injury accidents were due to individual failings (slips, lack of ability, lack of motivation and so on) and those arising from organizational shortcomings. Among the latter are accidents that could have been prevented by better training or instructions, better design, better construction, better maintenance and better methods of operation.

If both individual and organizational accidents have their roots in common systemic processes, then it could be argued that LTIF rates—or comparable personal injury statistics—are indicative of a system's vulnerability (or resistance) to organizational accidents. The number of personal injuries sustained in a given time period must surely be diagnostic of the 'health' of the system as a whole. Unfortunately, this is not so. The relationship is an asymmetrical one. An unusually high LTIF is almost certainly the consequence of a 'sick' system that could indeed be imminently liable to an organizational accident. But the reverse is not necessarily true. A low LTI rate (of the order of 2–5 per million manhours)—which is the case in many well run hazardous technologies—reveals very little about the likelihood of an organizational accident.

There are two parts to this problem. First, such low and often asymptotic LTIFs comprise more noise than signal. In one accounting period, for instance, a major part of all the recorded lost-time injuries could reflect only that a member of the administrative staff fell off a bicycle on an icy patch outside the office building and fractured a wrist. In the case of the oil and gas business, for example, what does this say about the integrity of an offshore installation? Nothing. And this brings us to the second problem. For the many concerned organizations who have reduced their personal injury events to what could be an irreducible minimum (given the ever-present hazards of the business), lost-time injuries are no longer any kind of indication of where the real dangers lurk. At this point, non-injury events—such as the number of leaks of combustible materials—are much more diagnostic of the integrity of the system as a whole. It is these precursor events, along with regular checks upon the quality of the underlying processes (see Chapter 7), that show where the high-potential risks are located.

Top-level commitment is a prerequisite for effective risk management, but it is not sufficient. Indeed, a blinkered commitment to the person model can be counterproductive. For example, a visitor to a drilling rig (or a comparable installation) is usually confronted by a large sign declaring that this site has had so many thousands of hours without a lost-time injury—quite often these are near-threshold values such as 999 970 hours. While such signs are clearly designed to motivate the workforce to maintain their safe working practices, they also convey two other, less helpful, messages. The first is that this is a safe place—which it is not (as a Texan driller once told me, 'There ain't a damn thing on this site that can't hurt you'). The second message is 'Woe betide the supervisor that *reports* a lost-time injury now'. As argued in Chapter 6, commitment needs to be combined with the other two 'Cs'—competence and cognisance. Together, they add up to an intelligent and informed awareness both of the varieties of risk and of the different ways to combat them.

Latent Conditions: The Universal Risk

Regardless of the personal hazards of the workplace, all organizations are vulnerable to latent conditions and to the breakdown of their defences. In this respect, the organizational model presented here, though derived largely from a study of physically damaging events, has relevance to all domains. But this is not always appreciated. In part, the problem is due to the close association in people's minds between safety and personal injury risks—if their system is not subject to physically harmful events, then they tend to be profoundly uninterested in anything that has 'safety' in the title. As this book has sought to demonstrate, however, both physical and economic disasters have common causal pathways. The same basic principles and countermeasures are as applicable to a bank or an insurance company as they are to chemical process plants and oil companies.

Has the Pendulum Swung too Far?

The earlier criticisms of the person model should not be taken as indicating that the organizational approach is problem-free. As discussed at several points throughout this book, the last 20 years have seen an ever-widening search for the origins of major accidents. Investigators and analysts have backtracked from the bad outcome, through the proximal unsafe acts (if identified), the workplace and organizational factors to the regulators and the system as a whole—and, in some cases, to the economic climate and the nature of the society at large. Figure 10.1 illustrates how some of these major events stand with regard to this extended causal fallout.

The question posed in the heading of this section is prompted by a suspicion that the pendulum may have swung too far in our present attempts to track down possible error and accident contributions that are widely separated in both time and place from the events themselves. The relative worth of the various causal categories can

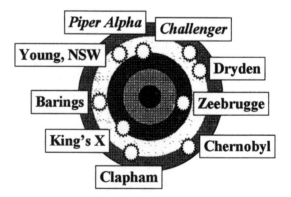

Figure 10.1   Map of the 'causal fallout' from recent organizational accidents
Moving from the centre outwards, the 'bull's eye' represents the front-line individuals, the 'inner' relates to the workplace, the next circle corresponds to organizational factors, the next to the regulators and the overall system, and the outer ring corresponds to societal factors.

be evaluated by reference to three questions central to the pursuit of system safety. To what extent does a consideration of individual, contextual, organizational, systemic and societal factors add value (see Figure 10.2):

- to our understanding of the causes of accidents and events?
- to our ability to predict the likelihood of future accidents and events?
- and, most importantly, to our remedial efforts to reduce their future occurrence?

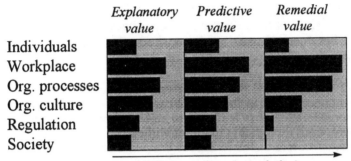

Relative values (speculative)

Figure 10.2 The relative (highly speculative) values of various types of possible causal factors for the explanatory, predictive and remedial goals
Individual factors alone have only a small to moderate value for all three goals and that, overall, workplace and organizational factors contribute the most added value. There are diminishing returns on more remote influences, particularly with regard to countermeasures and risk management.

While it is clear that the present situation represents a significant advance over knee-jerk 'human error' attributions, some concerns need to be expressed about the theoretical and the practical utility of this ever-spreading quest for contributing factors. We seem to have reached, or even exceeded, the point of diminishing returns, particularly when it comes to risk management. We also need to find some workable middle ground that acknowledges both the psychological and the contextual influences on human performance, as well as the interactions between active failures and the latent conditions that serve, on rare occasions, to breach the system's defences. Chapter 5, for instance,

presented a strong case for giving much closer attention to maintenance activities. Models of accident causation can only be judged by
the extent to which their applications enhance system safety. The economic and societal shortcomings, identified—for example—by Legasov
(see Chapter 1), are beyond the reach of system managers. From their
perspective, such problems are given and immutable, but our main
interest must be in the changeable and the controllable.

Some Problems with Latent Conditions

In earlier accounts,[15] these delayed-action 'time-bombs' were described as latent errors or latent failures. But in the causal sense, the
term 'condition' is much more appropriate. Hart and Honoré distinguished between 'causes' and 'conditions': causes are what 'made
the difference', 'Mere conditions', on the other hand, are:

> ... just those [things] that are present alike both in the case where
> accidents occur and in the normal case where they do not; and it is
> this consideration that leads us to reject them as the cause of the
> accident, even though it is true that without them the accident would
> not have occurred ... to cite factors that were present both in the case
> of the disaster and in normal functioning would explain nothing: such
> factors do not 'make the difference' ... .[16]

In the case of a fire, for example, a 'mere condition' would be the
oxygen in the air. In a railway accident, 'they will be such factors as
the normal speed and load and weight of the train and the routine
stopping and acceleration'. Although the latent conditions we have
considered in this book are not quite of this character, they are present
within the system regardless of whether or not an accident occurs.
All systems harbour latent conditions; an accident simply makes
them manifest. In short, their presence does not discriminate between normal states and bad events. Moreover, the extent to which
they are revealed will depend not so much upon the 'sickness' of the
system, but on the resources available to the investigator. The more
exhaustive the inquiry, the more latent conditions it will uncover.

So we are left with the following conclusion: only proximal events—
unsafe acts and local triggers—will determine whether or not an
accident occurs. If that is the case, why do we need to consider these
more distal factors at all? There are three compelling reasons why
latent conditions are important:

- They undoubtedly combine with local factors to breach defences. In many cases, they *are* weakened or absent defences.

- Resident 'pathogens' within the workplace and the organization can be identified and removed *before* the event.
- Local triggers and unsafe acts are hard to anticipate and some proximal factors are almost impossible to defend against (for example, forgetfulness, inattention, and the like).

Thus, despite their inherent problems, identifying and eliminating latent conditions proactively still offer the best routes to improving system 'fitness'. But it has to be a continuous process. As one problem is being addressed, others will spring up in its place. There are no final victories in the safety war.

The Price of Failure

In Chapter 5 we touched upon the costs of maintenance failures in commercial aviation. Since the issues discussed in this chapter were largely prompted by the oil industry, we will focus here upon the costs of accidental losses as they affect this domain. Another good reason for doing this is that the process of 'loss costing' has probably been carried further in this area than elsewhere.

Table 10.2   The financial losses incurred by major events in the petrochemical industry

Event	Country	Financial loss
Piper Alpha	United Kingdom	$2.5 billion*
Exxon Valdez	USA	$3.5 billion
Phillips 66 Pasadena	USA	$1.3 or $2.1 billion**
Sleipner A	Norway	$300 million
Saga 2/4-14	Norway	$250 million
La Mede	France	$260 million
Sodegaura	Japan	$171 million
Grangemouth	United Kingdom	$100 million
Croatzcoalcas	Mexico	$98 million
Pembroke	United Kingdom	$79 million
Dhaka	Bangladesh	$76 million
Ras Tanura	South Africa	$35 million

*   This does not include the incalculable cost of 167 fatalities. The HSE estimate the total cost of Piper Alpha (less these unquantifiables) at £2.066 billion. Moreover, the associated disruption to North Sea production knocked at least a percentage point off the growth rate of output during the affected period.
**   The higher estimate was given by Dan Stover of Brown & Root Energy Services.

Table 10.2 shows the estimated financial costs associated with a number of major events in the petrochemical domain. The data were provided by Eric Brandie, the Loss Prevention Manager for Chevron UK Ltd.[17]

A joint loss costing study was carried out by the HSE and Chevron covering 13 weeks typical operations on one of the latter's platforms in the North Sea. Although there were no serious incidents during that period, the costs accruing from a whole range of minor classified incidents amounted to £1 million. Projected on a field-wide annual basis, this rose to £4 million—equivalent to the shutdown of a platform for one day each week throughout the year.

In 1993, the HSE estimated that for every £1 of costs recoverable through insurance, another £5 to £50 are added to the final bill through a wide variety of other financial losses. Like an iceberg, for every visible pound (recovered from the insurers), there are up to 50 times that sum below the surface in indirect costs. These include:

- product and material damage
- plant damage
- building damage
- tool and equipment damage
- legal costs
- expenditure on emergency supplies
- clearing site
- production delays
- overtime working
- investigation time
- supervisors' time diverted
- cost of panels of inquiry
- clerical effort.

To these should also be added such intangibles as damage to the company's reputation (probably reflected in the share price), loss of business, recruitment difficulties and a general lowering of morale. Accidents do not only cost lives, they are also economically disastrous. Very few organizations can sustain these levels of financial loss.

The real question, of course, is not what safety costs us, but what it saves. This book has described how organizational accidents arise and has outlined practical measures for reducing the likelihood of their occurrence. But such accidents can afflict even the best run systems. It is therefore not enough simply to plan for their prevention, it is also essential to plan for post-accident business recovery in order to minimize these huge losses. Dan Stover[18] gives us an example of what an effective loss recovery procedure can achieve. The consequences of the massive Bishopsgate bombing in the City of London in 1993 were shared by many organizations. Among these, the Saudi International Bank, was open for business as usual the following Monday. This bank had in place a well conceived loss recovery programme that had been tested by the bombing in the nearby St Mary Axe. The total cost of this bombing was estimated at £1.5 billion.

Finally, a sobering thought: four out of five organizations suffering a major disaster without recovery procedures never reopen for business. Furthermore, 80 per cent of disaster recovery plans do not work the first time.[19] Recent studies[20] have shown that crisis-prepared organizations plan for at least five different types of crisis, and the crisis plans are closely linked to business recovery plans. In addition, such organizations have a flexible and adaptive structure and are low on both rationalizations and denial. As Denis Smith of Durham University Business School put it, 'Any denial of the mainstream nature of crisis management is a manifestation of a crisis-prone culture and, as such, becomes a suitable case for treatment'.[21]

The Last Word

In this chapter we have sought to reconcile three different approaches to safety management: the person model directed at reducing personal injury events; the engineering model focusing on the human–machine interface and system reliability; and the organizational model that deals with the integrity of defences and the broader systemic factors. Each has its own metrics and countermeasures. To the extent that any domain is exposed to the risks of both individual and organizational accidents, all of these models have their part to play in the overall safety management programme.

Such conflicts that exist arise mainly from the predominance of the person model in situations that demand a closer consideration of technical and systemic factors. Whereas the engineering and organizational models can be usefully applied to the reduction of individual accidents, the person model *alone* (and especially the mindset that goes with it) has very limited value in domains where the risks are mainly derived from the insidious accumulation of latent conditions and their rare conjunctions with local triggers to defeat the multi-layered defences. This does not mean, of course, that we should ignore the personal injury risks or the behaviour of individuals and teams. But it does mean that risk managers should be aware of the broader systemic origins of organizational accidents and of the variety of techniques now available to thwart their development. Effective risk management requires the application of different counter-

measures targeted at different levels of the system at the same time—
and all the time. It takes only one organizational accident to put an
end to all worries about the bottom line.

Notes

1   I am particularly grateful to Andy Pearce of Shell Expro, Richard Clark of BP
    Exploration and Rob Pinchbeck of Atlantic Power and Gas for making these
    and other valuable contacts with the professionals possible.
2   When a work-related injury results in some absence from work, usually in
    excess of 1–2 days.
3   D.A. Lucas, 'Wise men learn by others' harms, fools by their own: organiz-
    ational barriers to learning the lessons from major accidents', Paper given to
    the Safety & Reliability Society Symposium on 'Safety and Reliability in the
    '90's: Will past experience or prediction meet our needs?', 19–20 September,
    1990, Altrincham, Manchester. See also D.A. Lucas, 'Understanding the human
    factor in disasters', *Interdisciplinary Science Reviews*, **17**, 1992, pp. 185–90. See
    also Center for Chemical Process Safety, *Guidelines for Preventing Human Error
    in Process Safety*, (New York: Center for Chemical Process Safety of the Ameri-
    can Institute of Chemical Engineers, 1994), pp. 44–101.
4   F. Bird Jr., *Practical Loss Control Leadership*, (Loganville, GA: International Loss
    Control Institute, 1969).
5   E. Hollnagel, *Human Reliability Analysis: Context and Control*, (London: Aca-
    demic Press, 1993); Center for Chemical Process Safety, op. cit. A.I. Glendon
    and E.F. McKenna, *Human Safety and Risk Management*, (London: Chapman &
    Hall, 1995); B. Kirwan, *A Guide to Practical Human Reliability Assessment*, (Lon-
    don: Taylor & Francis, 1994), E.M. Dougherty Jr. and J.R. Fragola, *Human
    Reliability Analysis: A Systems Engineering Approach with Nuclear Power Plant
    Applications*, (New York: Wiley, 1988).
6   B. Turner, *Man-Made Disaster*, (London: Wykeham, 1978)—a second edition is
    currently being prepared by Dr Nick Pidgeon of the University of Wales,
    Bangor.
7   C. Perrow, *Normal Accidents: Living with High-Risk Technologies*, (New York:
    Basic Books, 1984).
8   *Report of the Royal Commission into the Crash on Mount Erebus, Antarctica, of a
    DC10 Aircraft Operated by Air New Zealand Limited*, (Wellington, 1981). See also
    G. Vette, *Impact Erebus*, (Wellington: Aviation Consultants Ltd, 1983) and S.
    McFarlane, *The Erebus Papers*, (Auckland: Avon Press, 1991).
9   D. Maurino, *et al.*, *Beyond Aviation Human Factors*, (Aldershot: Avebury, 1995),
    pp. 30–56.
10 Mr Justice V.P. Moshansky, Commission of Inquiry into the Ontario Crash at
    Dryden, Ontario. Final Report, Vol. 1 (Ottawa: Ministry of Supply and Ser-
    vices, 1992).
11 Mr Justice Sheen, *M.V. Herald of Free Enterprise. Report of Court No. 8074*, (De-
    partment of Transport, London: HMSO, 1987).
12 N. Cohen, 'Zeebrugge manslaughter prosecution collapses', *The Independent*, 20
    October, 1990.
13 H.L.A. Hart and A. Honoré, *Causation in the Law*, (2nd edn), (Oxford: The
    Clarendon Press, 1985), p. 43.
14 T.A. Kletz, *An Engineer's View of Human Error*, (Rugby: The Institution of Chemi-
    cal Engineers, 1985).

15 J. Reason, *Human Error*, (Cambridge: Cambridge University Press, 1990), ch. 7,

16 H. Hart and A. Honoré, op. cit., p. 34.

17 E.F. Brandie, 'The cost of accidents—an operator's view', Paper given to the 1996 Safety Conference 'People & Changes, Costs & Challenges', Institute of Petroleum, London, 26 September 1996.

18 D. Stover, The costs of failure. *Proceedings of the 1996 Safety Conference*, Institute of Petroleum, London, 26 September 1996.

19 Ibid.

20 I. Mitroff, T. Pauchant, M. Finney and C. Pearson, 'Do some organizations cause their own crises? The cultural profiles of crisis-prone vs. crisis-prepared organizations', *Industrial Crisis Quarterly*, 3, 1989, pp. 269–83.

21 D. Smith, 'Crisis management and strategic management', *Advances in Strategic Management*, 8, 1992, p. 268.

Index